T0140758

Novel Whispering Gallery Mode Microresonators and Waveguide Designs for Applications in Integrated Quantum Optics

D i s s e r t a t i o n

zur Erlangung des akademischen Grades
d o c t o r r e r u m n a t u r a l i u m
(Dr. rer. nat.)
im Fach Physik

eingereicht an der
Mathematisch-Naturwissenschaftlichen Fakultät
der Humboldt-Universität zu Berlin

von
Dipl.-Phys. Rico Henze, M.Sc.

Gutachter/innen:
1. Prof. Dr. Oliver Benson
2. Prof. Achim Peters, Ph.D.
3. Prof. John Donegan, Ph.D.

Tag der mündlichen Prüfung: 04.09.2014

Bibliographic information published by the Deutsche Nationalbibliothek

The Deutsche Nationalbibliothek lists this publication in the Deutsche Nationalbibliografie; detailed bibliographic data are available in the Internet at http://dnb.d-nb.de .

978-3-8325-3906-1

Logos Verlag Berlin GmbH
Comeniushof, Gubener Str. 47,
10243 Berlin
Tel.: +49 (0)30 42 85 10 90
Fax: +49 (0)30 42 85 10 92
INTERNET: http://www.logos-verlag.de

Abstract

This thesis deals with the investigation of novel resonator and waveguide designs with regard to their applicability in experimental quantum optics. As especially the light-matter interaction on smallest scales is in the foreground here, the at most compact and low-loss experimental setups are required for the observation of the occurring quantum effects. By using microintegration with optical and electronic components, highest levels of controllability and precision can be achieved.

In focus of this work are the so-called whispering gallery mode microresonators. These devices allow concentration and storage of light within highly confined volumes. The occurring energy densities are the basis of an effective interaction between the light field and its environment. Due to the extreme quality factors and necessary compactness of such systems, highly precise resonance tuning is required. Therefore, in this work microresonators based on various material systems are particularly examined for their tuning abilities.

The first system is formed by borosilicate-based hollow microsphere resonators. These allow using mechanical pressure control to achieve particularly wide tuning ranges of over several 100 GHz. Analytic models to calculate the geometric parameters and possible control ranges are specified for this system. For low temperature applications, the cryogenic tuning behavior is also measured and compared to room temperature readings. The studied resonators show excellent agreement with the made assumptions.

Further in this work, resonator systems produced with a plurality of lithographic processing techniques from semiconductor industry are examined. For post-production tuning silica-based disk resonators, a novel chemical etching method is presented. It is shown that the appropriate choice of a concentrated BHF etching solution allows a controlled and precise material removal for exactly tuning into a 10 GHz range. For the demonstration of fine-tuning, well-defined reference lines are approached by means of temperature control.

The applicability of such chip-based resonator systems for quantum optical investigations is demonstrated in an experiment by coupling single photon emitters to toroidal microcavities. Here, a novel method for the targeted manipulation of emitters is used and applied for showing the mutual coupling from corresponding scattering measurements. With this system, the applicability of such coupled photonic elements for applications in cold finger cryostats is also successfully demonstrated.

Furthermore, a novel resonator design based on a photopolymer functionalized with single photon emitters is presented. With this material, disk resonators and corresponding coupling elements are generated by laser direct writing. The structures are spectrally measured and calibrated. With Q factors around 10^4, they show optical properties comparable to silica. The simple manufacturing process and the material embedded emitters offer a very interesting alternative for quantum optical investigations.

In addition to the resonators, highly efficient coupling mechanisms for the excitation and readout of these structures are necessary for the later use of such systems. Therefore, precise resonator adapted waveguide designs are also required.

In this work, novel silica- and polymer-based waveguide designs were developed and examined for their optical properties. For these two systems, waveguide creation and quantum optical applicability is successfully demonstrated.

Plasmonic waveguides and alternative waveguide designs based on nitrides offer further interesting possibilities for quantum optics experiments. Besides these two waveguide designs, in this work also a corresponding coupler design is presented and investigated. It allows efficient coupling between for single-mode operation optimized dielectric and plasmonic structures. Thus, it admits the detailed studying of quantum optical phenomena also on single plasmons. In addition, this may enable the realization of plasmonic quantum networks.

Zusammenfassung

Diese Dissertation beschäftigt sich mit der Untersuchung neuartiger Resonator- und Wellenleitersysteme im Hinblick auf ihre Anwendbarkeit in der experimentellen Quantenoptik. Da hier insbesondere die Licht-Materie-Wechselwirkung auf kleinsten Skalen im Vordergrund steht, werden zur Beobachtung der auftretenden Quanteneffekte möglichst kompakte und verlustarme experimentelle Aufbauten benötigt. Durch Mikrointegration mit optischen und elektronischen Bauelementen kann dabei ein Höchstmaß an Kontrolle und Präzision erreicht werden.

Den Schwerpunkt dieser Arbeit bilden die sogenannten Flüstergalerieresonatoren. Diese ermöglichen eine Konzentration und Speicherung von Licht auf engstem Raum. Die dabei auftretenden Energiedichten bilden die Grundlage für eine effektive Wechselwirkung des Lichtfeldes mit seiner Umgebung. Aufgrund der extremen Resonatorgüten und der notwendigen Kompaktheit solcher Systeme ist eine möglichst präzise Abstimmung der Resonanzen notwendig. Im Rahmen dieser Arbeit wurden deshalb auf verschiedenen Materialsystemen beruhende Mikroresonatoren insbesondere auf deren spezifisches Abstimmverhalten hin untersucht.

Das erste System wird von auf Borosilikatglas basierenden Mikrohohlkugelresonatoren gebildet. Diese ermöglichen mittels mechanischer Druckregelung einen besonders weiten Abstimmbereich über mehrere 100 GHz. Für dieses System werden analytische Modelle zur Berechnung der geometrischen Parameter und des möglichen Regelbereichs angegeben. Für Tieftemperaturanwendungen wird zudem das kryogene Abstimmverhalten gemessen und mit Raumtemperaturwerten verglichen. Für die dabei untersuchten Resonatoren zeigt sich eine ausgezeichnete Übereinstimmung mit den gemachten Annahmen.

Weiterhin werden in dieser Arbeit mit verschiedenen lithographischen Methoden der Halbleiterprozessierung hergestellte Resonatorsysteme untersucht. Zur Abstimmung von auf Siliziumdioxid basierenden Scheibenresonatoren wird ein neuartiges chemisches Ätzverfahren zur Nachprozessierung vorgestellt. Es wird gezeigt, dass bei Wahl einer entsprechenden Konzentration eines BHF-Lösungsmittels ein kontrollierter und präziser Ätzabtrag für Abstimmungen bis auf 10 GHz genau möglich ist. Zur Demonstration einer Feinabstimmung werden dann mittels Temperaturregelung genau definierte Referenzlinien eingestellt.

Die Anwendung solcher Chip-basierten Resonatorsysteme für quantenoptische Untersuchungen wird in einem Experiment zur Kopplung von Einzelphotonenemittern an toroidale Mikrokavitäten demonstriert. Dabei wird ein neuartiges Verfahren zur gezielten Manipulation der Emitter genutzt und die gegenseitige Kopplung durch miteinander korrespondierende Streumessungen gezeigt. Mit diesem System wird zudem die Anwendbarkeit solcher gekoppelten Strukturen für Anwendungen in Kühlfingerkryostaten demonstriert.

Des Weiteren wird ein neuartiges Resonatorsystem auf Basis eines mit Einzelphotonenemittern funktionalisierten Photopolymers vorgestellt. In diesem Material werden Scheibenresonatoren und entsprechende Kopplungselemente mittels Laserdirektschreibens erzeugt. Die Strukturen werden spektral vermessen und kalibriert. Bei Gütefaktoren um die 10^4 zeigen sich mit Siliziumdioxid vergleichbare optische Eigenschaften. Durch das einfache Herstellungsverfahren und die direkt ins Material eingebetteten Emitter bilden solche Systeme eine äußerst interessante Alternative für quantenoptische Untersuchungen.

Für die spätere Nutzung solcher Systeme bedarf es neben den eigentlichen Resonatoren auch möglichst effizienter Kopplungsmechanismen zur Anregung und zum Auslesen der Strukturen. Daher sind entsprechend präzise an die Resonatoren angepasste Wellenleitersysteme notwendig.

Im Rahmen dieser Arbeit wurden neuartige Wellenleiter auf Siliziumdioxid- und Polymerbasis entwickelt und auf ihre optischen Eigenschaften hin untersucht. Für beide Materialsysteme konnten erfolgreich Wellenleitung und quantenoptische Anwendbarkeit demonstriert werden.

Plasmonenwellenleiter und alternative Wellenleitersysteme auf Basis von Nitriden bieten zudem weitere interessante Möglichkeiten für quantenoptische Experimente. Neben diesen beiden Wellenleitersystemen wird in dieser Arbeit auch ein entsprechendes Kopplerdesign vorgestellt und untersucht. Dieses gestattet eine effiziente Kopplung zwischen auf Einzelmodenbetrieb optimierten dielektrischen und plasmonischen Wellenleitern und erlaubt damit eine eingehende Untersuchung quantenoptischer Phänomene auch an einzelnen Plasmonen. Zudem können dadurch plasmonische Quantennetzwerke realisiert werden.

Contents

Chapter 1

Introduction

Most people will agree that the last decades were clearly ruled by the great advantages in microelectronics [1]. The achievements in industrial microprocessing, large scale production abilities, and an increasing circuit density of microchips had made devices possible which some years ago nobody has even dreamed of [2]. As all those devices became more and more connected to each other, another great technology began to rise. Optical transmission was the answer to the rapidly increasing amount of data produced by and for those interconnected devices [3]. Exchanging thereby electrons by photons allowed a tremendous increase in the maximum transmitted data rate by simultaneously decreasing the overall system costs. Photonics is actually coming more and more into these devices themselves [4]. The great demand for highest data rates even on shortest ranges brings the advantages of optical data transmission directly to the chip level [5]. And as microelectronics did years before, nowadays so-called microphotonics emerges and begins to rapidly develop [6].

Whenever data rates are going up, information density is increased. Often less and less physical content is used to transfer single bits of information [7]. Going to the absolute limit requires information encoded to smallest possible entities – depending on the used technology these are single quanta of light or matter. In microelectronics, transistors acting on single electrons are actually state of the art [8]. For microphotonics, a similar technology acting on single photons is still matter of research [9]–[11]. At these levels of integration, the quantum nature of light can not be neglected anymore. On the contrary, it can be exploited to create higher levels of functionality. The corresponding technology belongs to the quantum electronics [12] or quantum photonics regime [13]. In photonics, this requires a deep understanding of the physics of single photons and insights into the basic interactions of light and matter. These interactions may then be exploited for applications way beyond the scope of classical photonics or even quantum electronics. A very well-known future application are the so-called quantum computers which enable the use of multiple quantum states for parallel high-speed data processing [14]; another example are quantum encrypted coding schemes and data transfer protocols [15]. Before the limits of microphotonics can be reached, appropriate tools and methods for handling and controlling the quantized states of light must be developed and investigated. They can still base on conventional microoptics, but should take account for the highly fragile nature of photons. Hence, it seems straightforward to combine well-developed microelectronic technologies with modern photonics and incorporate both on single microchips [16]. This allows performing under well controlled and stabilized working conditions not possible with the bulky systems conventionally used for such applications.

This thesis deals with two main components of integrated microoptical systems [16]. These are novel resonator and waveguide designs which are investigated with regard to their applicability in experimental quantum optics where the properties of single photons and their interactions with matter are studied. For generating single photons, different non-classical light sources are applied [17]. In addition to quantum dots and single molecules, single defect color centers were used. For further studying, highly efficient methods of collecting, routing, isolating, and confining the emitted photons are required. As the observation photonic quantum properties is extremely challenging, experimental setups must fulfill very special requirements in terms of their spectral, geometrical, optical, and mechanical properties [16].

In focus of this work are so-called whispering gallery mode (WGM) microresonators. These types of resonators allow concentration and storage of light within highly confined volumes. By coupling non-classical light sources to such resonator designs, amplification effects may occur depending on the strength of the interaction [18]. They can lead to better photon yields and increased emission rates [19], or even allow multiple coherent interactions between the emitters and stored photons [20]. For observing such effects, the optical losses have to be minimized [21]. Furthermore, due to the extreme quality factors and the necessary compactness of such systems, precise tuning of the occurring resonances is required [22]. The highly accurate methods of the semiconductor industry are thus perfectly suited for fabrication. In this work, microresonators based on various material systems are particularly examined for their specific tuning abilities. As for most up-to-date quantum emitters low-temperature operation is required, the presented results mainly aim for compact chip-integrated solutions [23]. However, as tunability is an important factor in cavity quantum electrodynamics (CQED), an off-chip research system based on a novel pressure tunable microbubble design is also presented in addition to the fully chip-integrated designs. Different structures are characterized and evaluated for their applicability in quantum optics at room temperature and in cryogenic environments. Furthermore, a novel resonator design based on a photopolymer functionalized with single photon emitters is presented.

In addition to the resonators, highly efficient coupling mechanisms for the excitation and readout of the structures are necessary for the later use of such systems. Therefore, precise resonator adapted waveguide designs are also required for all practical applications [21]. They should provide minimum attenuation and allow good coupling to other optical components. For this purpose, two novel silica- and polymer-based waveguide designs were developed and intensively examined. Plasmonic waveguides in conducting media and alternative waveguide designs based on nitrides allow even smaller footprints and offer interesting new possibilities for the design of modern quantum optics experiments. Besides these two waveguide designs, in this work also a corresponding coupler design is presented and investigated. It allows an efficient coupling between for single-mode operation optimized dielectric and plasmonic structures. Thus, it admits the detailed studying of quantum optical phenomena also on single plasmons. In addition, this may enable the realization of plasmon-based quantum networks.

In **Chapter 2**, a general overview over the most important characteristics of WGM microresonators is presented. After shortly introducing their basic principles, the fundamental properties for characterizing these specific types of microresonators are summarized. For a detailed analysis of the modal structure inside such cavities, analytic solutions for disk-type and spherical microresonators are presented. The analysis of the WGM mode structure is further engrossed by showing numerical solutions of the problem. The numerical approach allows investigating more realistic designs and gives direct access to important system parameters required for analyzing the general coupling behavior between microresonators and arbitrary optical couplers. By means of the transfer matrix formalism, these results are exploited to describe the mutual influence of the coupling structure to different WGMs inside the cavity. The formalism is applied for single- and multimode interactions and presented in the last section of this chapter.

In **Chapter 3**, a first WGM resonator design for future applications in advanced quantum optics is investigated. This design is based on so-called microbubbles and allows continuous on-demand resonance tuning by simply applying variable aerostatic pressure to the inside of the cavity. The decoupling of the tuning mechanism from the outer environment and the large possible tuning ranges of these cavities are perfectly suited for applications in highly resonant quantum optics experiments. Furthermore, active resonance stabilization is suitable due to a fast material response time. After a brief introduction into the specific properties of such microresonator systems, the manufacturing process of these cavities is explained. For characterizing the produced microbubbles, a simple model for estimating the wall thickness from pre-production parameters is presented. After analytically modeling the pressure tuning properties of microbubbles, the theoretical results are compared to actual room temperature tuning data. For extending these experimental results into the cryogenic temperature range, a home-build cryostat system was developed. The measurement setup is explained and the results are compared to the room temperature readings. With these measurements, it was also possible to determine the temperature dependent resonance shift when cooling the resonators from room temperature to liquid nitrogen temperature. This dependency is an important requirement for engineering the resonance properties of coupled quantum systems.

In **Chapter 4**, chip-based microresonators are investigated for applications in quantum optics. After a short introduction to general lithography, the detailed process of fabricating chip-based high-Q silica disk and toroidal microresonators is presented. Both types of resonators are then characterized and experimentally investigated. As in quantum optics resonance tunability is an important aspect, well-known tuning techniques are summarized. Afterwards, a novel post-production etching scheme for the permanent tuning of on-chip microresonators is presented. The approach allows a selective resizing of the physical dimensions of the resonators with high accuracy. It is perfectly suited for experiments in a cryostat environment where the other tuning methods are not fully applicable. The required tuning ranges are determined from first experimental quantum optics experiments. Single photon emitting defect centers are attached to toroidal microresonators for an investigation of the optical properties in the combined quantum system. The measurements are performed under room temperature and cryogenic conditions. They successfully demonstrate the general applicability of chip-based microresonator designs for basic CQED research. In a cryostat, multiple emission lines of a single attached nanocrystal can be distinguished. Hence, this allows some sort of additional selective resonance tuning.

In **Chapter 5**, a novel design for building low-loss optical waveguides in silica-based material systems is investigated. This approach is compatible with the on-chip WGM resonator designs introduced in the previous chapter and enables the realization of fully integrated on-chip setups in experimental quantum optics. By combining resonators and waveguides at a common platform, highly efficient coupling structures can be realized. Furthermore, the design is also suitable for a wide range of other applications where low-loss air-cladded waveguides in silica and other material systems are required. After presenting the principal waveguide design rules, the corresponding lithographic fabrication method is explained. The optical properties of the waveguides are analyzed and the resulting mode structures are calculated.

In **Chapter 6**, another material system suitable for integrated resonator and waveguide designs is investigated. The system is based on photopolymerizable monomers often referred to as photopolymers. In such materials, waveguides and other optical structures can be produced by mainly two different types of direct laser writing. These are the standard two-photon absorption-based direct laser writing lithography (DLW) and the DLW lithography in diffusion-mediated photopolymers. These two writing schemes provide a very general approach for the realization of integrated optical components. They can be used as basic elements for future applications in integrated quantum optics. Two different applications using both methods are presented. First, a fully integrated quantum photonic circuit based on a novel functionalized photopolymer is demonstrated. The contained active quantum emitters are embedded and allow for freestanding polymer-based resonator and waveguide structures. Second, the direct fabrication of buried polymer waveguides is demonstrated with a novel DLW scheme. This method can be used for chip-based interconnections and allows flexible coupling between different optical components. This chapter starts with an introduction to important polymer material properties.

In **Chapter 7**, further approaches for integrated chip-based designs are investigated. It includes alternative material systems for the presented resonator and waveguide structures to allow simpler fabrication, and higher levels of integration and process control. For example, silicon nitride-based structures are discussed as replacement for silica-based systems. Another design approach is based on so-called nanoplasmonics. These are applications where collective electromagnetic oscillations in conductive materials are applied for waveguiding. In nanoplasmonics, the classical diffraction limit of light can be beaten and ultra-high levels of integration are achievable. As these structures sustain the quantum properties of light, they can be further applied for using in quantum plasmonics. Hence, the fundamental physics of plasmonic waveguides is discussed. Furthermore, a novel coupler design for the efficient coupling between dielectric and plasmonic waveguides is introduced. The coupler allows bridging the gap between dielectrics and plasmonics, and enables the interconnection of individual nanoplasmonic elements. For quantum optics applications, the high field enhancement in the vicinity of plasmonic structures is important. This highly enhances the interaction of coupled quantum systems and allows increasing the total photon count rates. By mainly using simple and efficient processing techniques, the presented coupler design even allows the realization of highly integrated plasmonic networks for future quantum plasmonics applications.

[1] T. Forester, *The Microelectronics Revolution*. Cambridge: MIT Press, 1981.

[2] J. D. Meindl, "Low power microelectronics: retrospect and prospect," in *Proceedings of IEEE*, 1995, vol. 83, no. 4, pp. 619–635.

[3] E. Bonetto, L. Chiaraviglio, D. Cuda, G. A. Gavilanes Castillo, and F. Neri, "Optical technologies can improve the energy efficiency of networks," in *35th European Conference on Optical Communication (ECOC)*, 2009, pp. 1–4.

[4] E. Suhir, "Microelectronics and photonics — the future," *Microelectronics J.*, vol. 31, no. 11–12, pp. 839–851, 2000.

[5] N. Kirman, M. Kirman, R. K. Dokania, J. F. Martinez, A. B. Apsel, M. A. Watkins, and D. H. Albonesi, "Leveraging Optical Technology in Future Bus-based Chip Multiprocessors," in *Proceedings of the 39th Annual IEEE/ACM International Symposium on Microarchitecture (MICRO)*, 2006, pp. 492–503.

[6] M. Smit, J. van der Tol, and M. Hill, "Moore's law in photonics," *Laser Photon. Rev.*, vol. 6, no. 1, pp. 1–13, 2012.

[7] S. Das Sarma, "Spintronics," *Am. Sci.*, vol. 89, no. 6, pp. 516–523, 2001.

[8] Z. A. K. Durrani, *Single-Electron Devices and Circuits in Silicon*. London: Imperial College Press, 2009.

[9] H. Takesue, N. Matsuda, E. Kuramochi, W. J. Munro, and M. Notomi, "An on-chip coupled resonator optical waveguide single-photon buffer," *Nat. Commun.*, vol. 4, no. 2725, pp. 1–7, 2013.

[10] L. Neumeier, M. Leib, and M. J. Hartmann, "Single-Photon Transistor in Circuit Quantum Electrodynamics," *Phys. Rev. Lett.*, vol. 111, no. 6, p. 063601, 2013.

[11] R. H. Hadfield, "Single-photon detectors for optical quantum information applications," *Nat. Photonics*, vol. 3, no. 12, pp. 696–705, 2009.

[12] F. A. Zwanenburg, A. S. Dzurak, A. Morello, M. Y. Simmons, L. C. L. Hollenberg, G. Klimeck, S. Rogge, S. N. Coppersmith, and M. A. Eriksson, "Silicon quantum electronics," *Rev. Mod. Phys.*, vol. 85, no. 3, pp. 961–1019, 2013.

[13] M. Lobino and J. L. O'Brien, "Quantum photonics: Entangled photons on a chip," *Nature*, vol. 469, no. 7328, pp. 43–44, 2011.

[14] O. Hosten, M. T. Rakher, J. T. Barreiro, N. A. Peters, and P. G. Kwiat, "Counterfactual quantum computation through quantum interrogation," *Nature*, vol. 439, no. 7079, pp. 949–952, 2006.

[15] A. Beveratos, R. Brouri, T. Gacoin, A. Villing, J.-P. Poizat, and P. Grangier, "Single Photon Quantum Cryptography," *Phys. Rev. Lett.*, vol. 89, no. 18, p. 187901, 2002.

[16] A. Politi, J. Matthews, M. G. Thompson, and J. L. O'Brien, "Integrated Quantum Photonics," *J. Sel. Top. Quantum Electron.*, vol. 15, no. 6, pp. 1673–1684, 2009.

[17] M. D. Eisaman, J. Fan, A. Migdall, and S. V Polyakov, "Invited review article: Single-photon sources and detectors," *Rev. Sci. Instrum.*, vol. 82, no. 7, p. 071101, 2011.

[18] D. Englund, A. Faraon, I. Fushman, N. Stoltz, P. Petroff, and J. Vucković, "Controlling cavity reflectivity with a single quantum dot," *Nature*, vol. 450, no. 7171, pp. 857–861, 2007.

[19] J. Wolters, A. W. Schell, G. Kewes, N. Nüsse, M. Schoengen, H. Döscher, T. Hannappel, B. Löchel, M. Barth, and O. Benson, "Enhancement of the zero phonon line emission from a single nitrogen vacancy center in a nanodiamond via coupling to a photonic crystal cavity," *Appl. Phys. Lett.*, vol. 97, no. 14, p. 141108, 2010.

[20] T. Aoki, B. Dayan, E. Wilcut, W. P. Bowen, A. S. Parkins, T. J. Kippenberg, K. J. Vahala, and H. J. Kimble, "Observation of strong coupling between one atom and a monolithic microresonator," *Nature*, vol. 443, no. 7112, pp. 671–674, 2006.

[21] T. J. Kippenberg, S. M. Spillane, D. K. Armani, and K. J. Vahala, "Fabrication and coupling to planar high-Q silica disk microcavities," *Appl. Phys. Lett.*, vol. 83, no. 4, pp. 797–799, 2003.

[22] W. von Klitzing, R. Long, V. S. Ilchenko, J. Hare, and V. Lefèvre-Seguin, "Frequency tuning of the whispering-gallery modes of silica microspheres for cavity quantum electrodynamics and spectroscopy," *Opt. Lett.*, vol. 26, no. 3, pp. 166–168, 2001.

[23] J. M. Ward, R. Henze, M. Gregor, C. Pyrlik, and A. Wicht, "Integrated Whispering-Gallery Mode Resonators for Fundamental Physics and Sensing Applications," in *Proceedings of SPIE*, 2012, vol. 8236, p. 82361C.

Chapter 2

Whispering Gallery Mode Resonators

2.1 Introduction

Optical resonators are key elements in modern optics. They are widely used as resonant cavities in lasers [1], as etalons for optical filters [2], as precise measuring tools [3], and for non-linear optics [4]. Due to their large size and heavy weight, conventional bulk optical components show serious limitations in their stability and the possible alignment. Hence, during the last decades integrated resonator designs have emerged as interesting alternatives. These compact structures are ranging from simple Fabry-Pérot cavities [5] over various Bragg [6] and distributed feedback elements [7] to the group of racetrack [8] or ring microresonators [9]. The last two of these elements distinguish themselves from the others by exhibiting a special type of circumferential resonance. They do not cause the light to undergo any hard reflections and allow building up a continuous unidirectional light propagation. This is the reason why in these types of cavities extremely long photon storage times and ultra-low losses can be observed. Within such resonator designs, special types of morphology dependent resonances, so-called whispering gallery modes (WGMs), are created. In WGMs, a beam of light is confined to a dielectric structure with circular or elliptical symmetry by the process of continuous, in principal lossless, total internal reflection. This reflection takes place directly at the interface between the outer rim of the resonator and its environment. Such systems exhibit propagational modes showing a strong similarity to an acoustic phenomenon which can be observed at the dome of the St. Paul's cathedral in London (see FIG. 2.1a). This phenomenon was first mathematically described and investigated in 1910 by Lord Rayleigh [10]. In this circular structure, 32 m in diameter, a whisper against the inner wall of the gallery can be heard around the full circumference by any listener also placing his ear against the wall. But although knowing this effect for more than 100 years now, the possible applications of the optical analogue were recognized just recently.

For example, microresonators showing WGMs for electromagnetic waves are unique tools for studying quantum electrodynamics. Due to their ultra-low losses, WGM microresonators are used wherever narrow linewidths, a high degree of wavelength selectivity, or ultra-long photon storage times are the crucial parameters. Thus, these resonators are of high interest for the telecommunication industry [11], e.g., as wavelength selective elements in add-drop multiplexers [12], as optical switches [13], or in signal delay lines [14]. But also outside telecommunication, the unique properties of micro-integrated resonator designs exhibit interesting alternatives compared to classical macroscopic approaches. Their small mode volumes in combination with high surface electric field strengths make these types of resonators very appealing for a wide range of applications. For instance, in FIG. 2.1b the schematic drawing of an alignment-free fiber taper coupled microsphere resonator system for optical gas sensing applications is presented [15]. Passive systems allow biological or chemical trace detection and analysis [16]. Due to their high intrinsic environmental sensitivity, such microcavities are also suited for the detection of single molecules [17], [18]. Within the field of microfluidic bio sensing, the surface of the resonator can be additionally functionalized to increase the detection efficiency from adding a sensing selectivity to specific molecules [19].

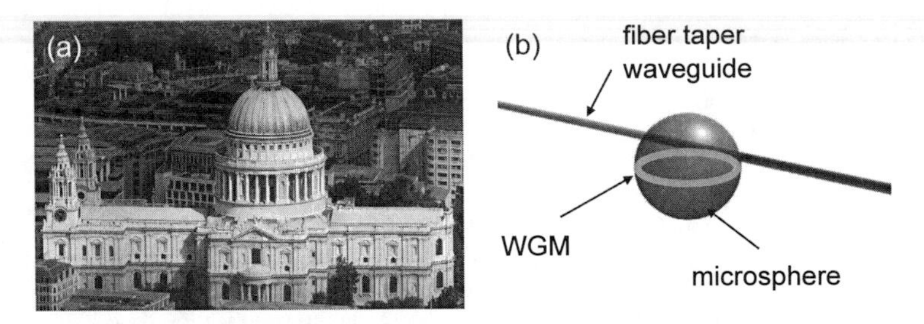

FIG. 2.1. Whispering galleries (from [20]). (a) Dome of the St. Paul's Cathedral in London. (b) Fiber taper coupled to a microsphere resonator. The system can be used for gas sensing applications.

In WGM resonators, light can be confined within smallest volumes. The high resulting field strengths allow nonlinear effects to be observed even in materials where these phenomena are usually not prominent [21]. A possible utilization of this effect is the generation of optical frequency combs for metrology applications and high precision frequency measurements. Due to an enhanced non-linear Kerr interaction between a continuously pumping laser field and the optical modes inside a monolithic WGM resonator, an equidistant resonance frequency spacing over a range of 500 nm could be successfully demonstrated [22]. Another interesting application of integrated WGM resonators is fundamental research on light-matter interactions at the nanometer scale. For instance, the coupling between optical and mechanical degrees of freedom by radiation pressure could already be demonstrated within such resonant structures [23]. Also fundamental research in cavity quantum electrodynamics (CQED), e.g., the observation of strong coupling between a microdisk and a quantum dot, could be realized [24].

In this chapter, a general overview over the most important aspects of WGM resonators is presented. After a short introduction into the basic principles of WGMs, fundamental properties for characterizing these kinds of microresonators are summarized. For a detailed understanding of the internal mode structure inside these cavities, analytic solutions for two different resonator designs, i.e., the disk-type and the spherical microresonator, are presented. The analysis of the WGM mode structure is further engrossed by showing a numerical design approach in addition to these simple analytic models. The presented software-based solution allows the investigation of more realistic structures and gives direct access to important parameters required for a detailed analysis of the general coupling properties between the cavities and arbitrary coupler designs. By means of the transfer matrix formalism, these parameters can be used to exploit the mutual influence between different fields inside these structures. This formalism is presented in the last section of this chapter for as well single- and multimode interactions.

2.2 Characteristic Properties of Microresonators

The properties of WGM microresonators are widely studied for a variety of different designs. For dielectric spheres and microspheres, the corresponding analytic models can be fully solved and the appearing resonances have been analyzed in detail [25]. For other designs, e.g., toroidal or bottle microresonators, at least perturbatively derived solutions can be found to describe their internal mode structure [26]. As all WGM resonators exhibit some sort of circular symmetry, in which the light is confined and constantly refocused, there are some fundamental properties shared by all of these resonators. They are independent on a specific design and can be used to describe and compare different resonator shapes. In the following section, these main properties are introduced to give the physical background for understanding the effects which are involved by working with these cavities. For this purpose, the exact definitions of the quality factor (Q factor), the free spectral range (FSR), and the spectral linewidth of a resonance are given.

2.2.1 Resonance Condition

A very simple and intuitive explanation of WGMs can be found with a graphical interpretation of the observed phenomenon. The electromagnetic field inside a spherical or elliptical microresonator can be seen as a ray of light which is constantly reflected along the inner surface of the resonator [27] due to total internal reflection (TIR) [28]. After one or more cavity roundtrips, the light, which is now interpreted as confined wave, can interfere with itself. If the cavity roundtrip length is an integer multiple of the wavelength of the circulating light, constructive interference occurs and is observed as distinct WGMs at elevated wavelengths

$$\lambda_{res} \approx \frac{2\pi \ r_{eff} n_{eff}}{m}. \tag{2.1}$$

Here, r_{eff} and n_{eff} are the effective radius and effective refractive index of the resonator, and m is the azimuthal mode number. This is an integer number corresponding to the number of intensity maxima around the circumference. The given approximation is strictly valid only for large azimuthal mode numbers $m >> 1$. In FIG. 3.2, two examples for the graphical interpretation of WGMs are given.

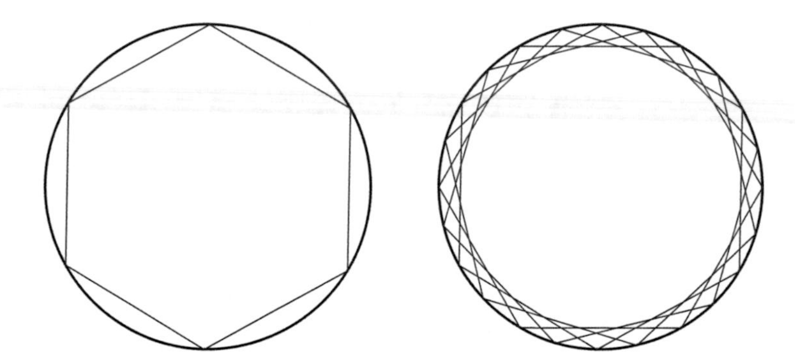

FIG. 2.2. Examples for the graphical interpretation of WGMs based on total internal reflection of rays. The left figure illustrates interference after one roundtrip with a low azimuthal mode number $m = 6$. In the right figure, a condition with a higher mode number $m = 24$ is shown. The constructive interference occurs here only after several roundtrips.

2.2.2 Quality Factor

An important parameter for resonant systems is their quality factor or Q factor. This parameter describes how strongly the oscillations in these systems are damped and thus how large the losses in the resonators are. Thereby, higher Q factors correspond to lower energy loss rates in relation to the total energy of the systems. The Q factor is commonly defined as 2π times the resonant frequency ν_{res} times the ratio between the stored energy U and the energy loss rate $-dU/dt$ as [29]

$$Q \equiv 2\pi \frac{stored\ energy}{energy\ loss\ rate} = 2\pi\nu_{res} \frac{U}{-dU/dt} = 2\pi\nu_{res}\tau\ . \tag{2.2}$$

For optical resonators, the resulting time constant τ describes the effective lifetime of a photon inside the cavity. This is in full analogy to the lifetime of an excited atom decaying by spontaneous emission. By solving the given differential equation for U, the time-dependence of the stored energy $U(t)$ can be calculated by

$$U(t) = U_0 e^{-2\pi\nu_{res}t/Q} = U_0 e^{-t/\tau}\ . \tag{2.3}$$

By means of a Fourier transformation, the spectral form of the resonance can be calculated. The resulting profile shows a Lorentzian lineshape with the maximum at $2\pi\nu_{res}$ and a full width at half maximum (FWHM) value of $\delta\nu_{res} = \nu_{res}/Q$. This leads to an alternative definition for the Q factor based on the spectral linewidth $\delta\lambda$ of a resonance [30]. For systems with $\delta\lambda << \lambda$ this can be written as

$$Q = \frac{\nu_{res}}{\delta\nu} \approx \frac{\delta\lambda}{\lambda_{res}}\ . \tag{2.4}$$

In a real resonator, several different loss mechanisms are simultaneously contributing to the total system loss. They are responsible for the total Q factor of the resonator [31]. If the respective contributions are small, then a separated analysis of the different loss channels is possible [32]. As each of these contributions is individually broadening the resonance linewidth, the total Q factor can be given as

$$Q^{-1} = \sum_i Q_i^{-1} = Q_0^{-1} + Q_{ext}^{-1}\ . \tag{2.5}$$

The main contributions to the total Q factor can be separated into two groups which correspond to internal and external loss channels [33]. These groups are the intrinsic Q factor Q_0 and the extrinsic Q factor Q_{ext}.

For the intrinsic Q_0 several loss components have to be considered. These are contributions due to material absorption inside the resonator Q_{abs}, radiation losses due to the curved surface of the resonator Q_{rad}, scattering losses caused by roughness and possible imperfections of the surface Q_{ss}, and induced losses due to material or dielectric disturbances Q_{cont}.

The extrinsic Q_{ext} is mainly contributed from losses induced by evanescently coupling the resonator to external devices, e.g., with a fiber taper or an optical prism. The coupler disturbs the homogeneity of the evanescent field of a WGM. It produces a scattering center for the circulating wave and limits the lifetime of the mode.

Depending on the specific resonator design, often one of the given loss components is highly dominating and thus limiting the achievable Q factor of the whole system. Typical values for silica microresonators in the visible spectral range and for radii around 25 μm are in the range of 10^5 to 10^6. Higher Q factors can be observed in silica resonator designs with a surface tension induced thermal reflow, e.g., in microtoroids or microspheres. For chip-based designs, ultimate Q factors in the range of 10^9 have recently been shown in oversized wedge resonators [34]. In the following paragraphs, two main loss contributions, which often limit the achievable Q factors in practice, are quantified.

Scattering Loss

For WGMs in directly etched microresonators, the optical losses induced by surface roughness and other imperfections are normally the most dominating factor [31]. This is due to the lithographic structuring and subsequent etching processes which typically result in an average surface roughness of R_a = 1.5 nm for silica-based systems [35]. In microresonators with a thermal reflow, the surface tension of the melted structure redefines the final shape of the cavity. By this, in silica microspheres and toroidal microresonators the surface roughness can be reduced to R_a' = 0.2 nm thus giving the higher Q factors achievable with such kind of resonators [36]. The scattering dependent component of the Q factor in a microdisk can be described with a modulated dielectric constant as [31]

$$Q_{ss} = \frac{3\lambda_{res}^3}{8\pi^{7/2}\zeta}\left(\frac{V_r}{V_s^2}\right). \tag{2.6}$$

Here, V_s is the volume of a typical scattering center whereas V_r is the physical volume of the resonator. ζ is a function of the refractive indices of the resonator n_r and the environment n_0, and the effective refractive index of the WGM n_{eff}. This functional term is given by

$$\zeta = \frac{n_0 n_{eff}^2 (n_r^2 - n_0^2)^2}{n_r^2 (n_{eff}^2 - n_0^2)}. \tag{2.7}$$

As the scattering loss is direct proportional to the ratio between the volumes of the resonator and the scattering defects, it can be deduced that for a constant surface roughness this Q factor component increases linearly with the geometrical size of the structure. If scattering defects are the main loss channel in a microresonator, then highest Q factors can be observed for large resonator geometries [34]. In microdisk resonators, the measured Q factors for TM modes are significantly higher compared to TE modes [37]. This can be explained by the inherently lower sensitivity of TM modes for such surface related defects.

Material Absorption

Another important loss factor is caused by absorption inside the resonator material. For a rough estimate, the corresponding Q factor component Q_{abs} can be approximated with the volume attenuation coefficient α by [38]

$$Q_{abs} \approx \frac{2\pi\, n_{eff}}{\alpha\lambda_{res}}. \tag{2.8}$$

In the wavelength range between 200 nm and 6.7 μm, this gives $Q_{abs} > 10^9$ for silica microresonators and thus material absorption can be neglected as limiting factor in this material [39]. In polymer-based resonator systems, the attenuation coefficients are normally around 0.1 $dBcm^{-1}$ [40]. This is in the same range as for silica-based materials and thus equivalent values for Q_{abs} can be achieved with polymers.

2.2.3 Mode Volume

While the Q factor describes the temporal confinement of a specific resonator mode, the mode volume V_m corresponds to the spatial extent of a mode. This parameter is especially interesting for experiments dealing with nonlinear effects or in CQED applications. Thereby, smaller mode volumes correspond to higher optical energy densities. This can dramatically enhance the observed nonlinearities of the resonator material. The mode volume is commonly defined as an equivalent volume with a homogeneous intensity distribution by the ratio between the total energy in the mode and its intensity maximum as [41]

$$V_m = \frac{\int_V \varepsilon(\mathbf{r})|E(\mathbf{r})|^2 d\mathbf{r}}{\max[\varepsilon(\mathbf{r})|E(\mathbf{r})|^2]}. \tag{2.9}$$

The mode volume can also be defined by another equation based on the effective nonlinearity of the mode [42]. This expression can be used to solve for an analytic solutions of the mode volume and is given by

$$V_m \approx \frac{\left(\int_V \varepsilon(\mathbf{r})|E(\mathbf{r})|^2 d\mathbf{r}\right)^2}{\int_V |E(\mathbf{r})|^4 d\mathbf{r}} \tag{2.10}$$

with $\varepsilon(\mathbf{r}) = n^2$ for the local dielectric constant of the material inside the resonator and $\varepsilon(\mathbf{r}) = 1$ for the dielectric constant of the environment. $E(\mathbf{r})$ is the amplitude of the electric field of the mode. Both values are taken at position **r** and integrated over a volume V. The mode volumes of microresonators with 50 μm diameter are typically in the range of 500 μm^3.

2.2.4 Free Spectral Range

The free spectral range (FSR) is a measure for the frequency or wavelength spacing between consecutive modes around a specific wavelength. These modes have common axial and radial mode numbers and identical polarization states, but differ in their azimuthal mode number by exactly one order. For large microresonators with $r >> \lambda$ the interpretive ray picture allows to conclude from Eq. (2.1) that the FSR of the resonators is approximately given by

$$\Delta\lambda_{FSR} = \frac{\lambda_{res}^2}{2\pi\, n_{eff} r_{eff} + \lambda_0} \approx \frac{\lambda_{res}^2}{2\pi\, n_{eff} r_{eff}}. \tag{2.11}$$

The FSR allows a simple measurement of the geometrical size of a microresonator. It also enables to explicitly map the observed mode families [43]. Between TE and TM modes of the same order, a difference in their specific resonance frequencies and FSRs can be observed. In TM modes, the FSR is slightly shifted to larger wavelengths due to the Goos-Hänchen effect. Both polarization modes experience different relative phase shifts on TIR [44]. For microresonators with 50 μm in diameter, a typical FSR is in the range of 5 nm at wavelengths around λ_{res} = 780 nm.

2.2.5 Finesse

The spectral linewidth of a resonance and the FSR of a resonator can be combined to define a dimensionless parameter called finesse. It is defined as the ratio between both values and describes how well two successive modes, which define the FSR of a system, can be separated in terms of their spectral linewidth (see FIG. 2.3). Thus, based on the Eqs. (2.1) and (2.2) for a specific wavelength the finesse of a resonant system can be calculated by

$$F = \frac{\Delta\lambda_{FSR}}{\delta\lambda} = \frac{\lambda_{res} Q}{2\pi\, n_{eff} r_{eff}}. \tag{2.12}$$

The finesse can be physically interpreted as 2π times the number of roundtrips inside the resonator before the initial intensity has decayed by a factor of e [30]. For most microresonator applications, small mode volumes in combination with high Q factors are required. This condition leads to high values for the finesse in such resonator systems. In a typical microresonator, finesses between 10^4 and 10^5 can be observed.

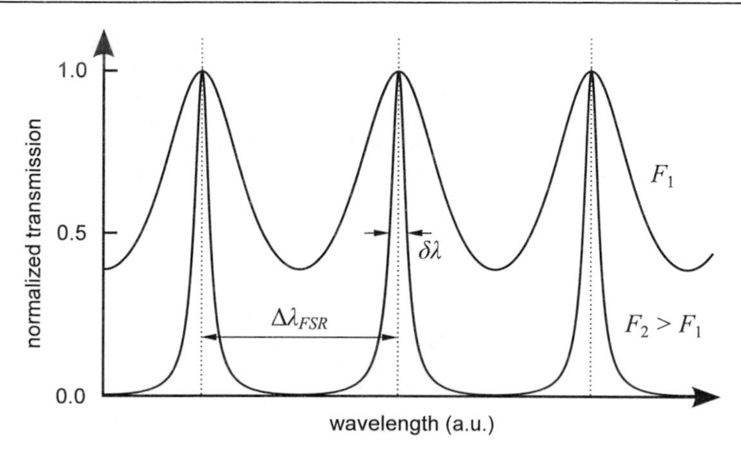

FIG. 2.3. Free spectral range $\Delta\lambda_{FSR}$ and spectral linewidth $\delta\lambda$ of a generalized resonance with different finesses F.

2.3 Analytic Model

Optical WGMs can occur wherever light is guided along an entirely closed spherical or cylindrical surface. This surface builds the interface between the resonator material and its local environment. If the resonator material shows the higher index of refraction, then TIR can occur when a sufficiently large refractive index contrast between the two regions on both sides of the interface is given. The circulating light is then trapped and confined to the inner surface of that interface. Under these conditions, highest Q factors and nearly lossless propagation can be observed.

In mathematical terms, WGMs can be described as solutions of the Helmholtz equation in a curved coordinate system. Even for the most simple geometries, e.g., spherical and disk resonators, fully analytic vector solutions are quite difficult to derive. Instead, approximations have to be made for reducing the complexity of the system. Often perturbative approaches or numerical simulations have to be applied. However, it can be shown that for most geometries exhibiting WGM resonances nearly all of the optical energy is contained either in the E or H fields perpendicular to the direction of propagation and that the corresponding state of polarization stays constant during energy circulation [45]. Especially for larger geometries, a simple scalar approximation becomes sufficient to describe the optical modes inside the system and a separable representation of the Helmholtz equation can be used instead of a fully rigorous vector model.

The following section describes a general ansatz for a purely analytic approach and the scalar solution of the Helmholtz equation for the simple case of a disk resonator is presented. Furthermore, based on the same ansatz also the solution of the Helmholtz equation in spherical coordinates is given. This solution is the basis to describe the optical modes in microbubble and hollow microsphere resonators and shows the generality of the presented analytic approach.

2.3.1 Optical Modes of Disk-Type Microresonators

The optical fields of a WGM have to obey Maxwell's equations inside and outside the resonator. In linear non-dispersive media without free charges and currents, the electric and magnetic fields can be assumed to exhibit a sinusoidal time-dependence of the form $e^{i\omega t}$. In this case, the time-independent Maxwell's equations can be used and the occurring fields are described by [29]

$$\begin{aligned} \nabla \cdot \mathbf{D}(\mathbf{r}) &= 0 \qquad & \nabla \times \mathbf{E}(\mathbf{r}) &= i\omega\, \mathbf{B}(\mathbf{r}) \\ \nabla \cdot \mathbf{B}(\mathbf{r}) &= 0 \qquad & \nabla \times \mathbf{H}(\mathbf{r}) &= -i\omega\, \mathbf{D}(\mathbf{r}) \end{aligned} \tag{2.13}$$

with $\mathbf{D}(\mathbf{r}) = \varepsilon(\mathbf{r})\mathbf{E}(\mathbf{r})$ and $\mathbf{H}(\mathbf{r}) = 1/\mu_0\, \mathbf{B}(\mathbf{r})$. $\varepsilon(\mathbf{r})$ is the local electric permittivity and μ_0 is the magnetic permeability of free space. From Eq. (2.13), the general Helmholtz equation for the different fields can be formulated as [46]

$$\nabla^2 \mathbf{\Psi}(\mathbf{r}) + k_0^2 n^2(\mathbf{r}) \mathbf{\Psi}(\mathbf{r}) = 0\,. \tag{2.14}$$

Here, $\mathbf{\Psi}(\mathbf{r}) = \{\mathbf{E}(\mathbf{r}), \mathbf{H}(\mathbf{r})\}$ corresponds the different components of the local electric and magnetic fields at vector position $\mathbf{r}$, $k_0 = \omega/c$ is the actual wave number in free space, and $n(\mathbf{r}) = \sqrt{\varepsilon(\mathbf{r})\mu(\mathbf{r})}$ describes the local refractive index based on the electric permittivity and magnetic permeability.

Helmholtz equation in cylindrical coordinates

Microdisk resonators possess a perfect cylindrical symmetry. Therefore, the natural coordinate system for the description of WGMs inside such structures are cylindrical coordinates with $\mathbf{r} = (\rho, \varphi, z)$. By transforming Eq. (2.14), the Helmholtz equation can be written as [47]

$$\left[\frac{\partial^2}{\partial \rho^2} + \frac{1}{\rho}\frac{\partial}{\partial \rho} + \frac{1}{\rho^2}\frac{\partial^2}{\partial \phi^2} + \frac{\partial^2}{\partial z^2} + k_0^2 n(r)^2 \right] \mathbf{\Psi}(\mathbf{r}) \;=\; 0\,. \tag{2.15}$$

When assuming microresonators with a sufficiently thin disk thickness $d \approx \lambda$ and high refractive index contrast between the resonator material and its environment, then the problem can be reduced into two dimensions. The circulating waves are fully guided within the resonator plane and thus two dominant polarization modes are occurring. These are the so-called transverse electric (TE) and transverse magnetic (TM) modes of propagation. The notation describes the orientation of the different field components within the plane. In a TE mode, the **E** field is oriented parallel to the disk, whereas the corresponding **H** field is perpendicular and shows continuous behavior across the side of the resonator (see FIG. 2.4). In a TM mode, the **E** field is oriented perpendicular to the disk. The field component is continuous across the resonator side. By considering this relations, Eq. (2.15) becomes scalar in *z*-direction and the vector function $\mathbf{\Psi}(\mathbf{r})$ is reduced to $\Psi(\mathbf{r}) = \{E_z(\mathbf{r}), H_z(\mathbf{r})\}$ for the TE/TM modes.

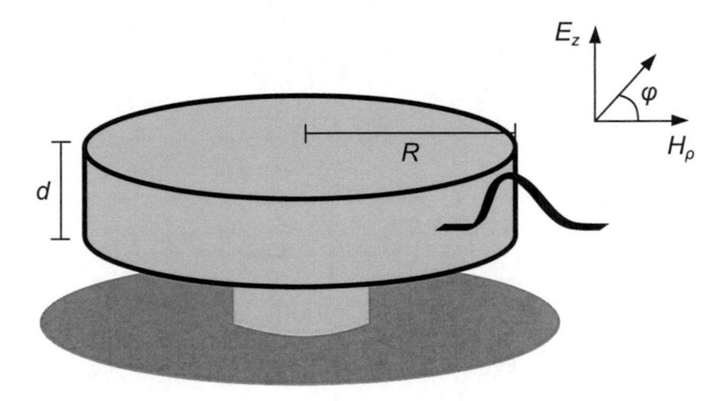

FIG. 2.4. Field configuration for the TM modes in a WGM disk resonator. The structure is geometrically characterized by disk thickness *d* and radius *R*.

For solving Eq. (2.15), the method of variables separation can be employed. By using $\Psi(\mathbf{r}) = U(\rho)\Phi(\varphi)Z(z)$, splitting into separate equations for the radial, azimuthal, and axial directions is feasible. The three resulting equations are [47]

$$\left[\frac{\partial^2}{\partial\rho^2} + \frac{1}{\rho}\frac{\partial}{\partial\rho} + \left(k_0^2 {n_{eff}}^2 - \frac{m^2}{\rho^2}\right)\right] U(\rho) = 0 \tag{2.16}$$

$$\left[\frac{\partial^2}{\partial\phi^2} + m^2\right] \Phi(\phi) = 0 \tag{2.17}$$

$$\left[\frac{\partial^2}{\partial z^2} + k_0^2\left(n^2(z) - n_{eff}^2\right)\right] Z(z) = 0 \tag{2.18}$$

with n_{eff} for the effective refractive index of the mode. The decoupling requires the introduction of an integer parameter m which connects the different solutions to each other. This parameter directly corresponds to the previously introduced azimuthal mode number m of a specific WGM resonance. It basically describes the number of intensity oscillations in the field around the full circumference of the cavity. The azimuthal mode number must be integer to obey the required resonance condition for the WGM inside the resonator.

For the axial direction, the analytic solution of Eq. (2.18) can be found in analogy to a symmetric slab waveguide with thickness d [30]. The equivalent solution is well known and describes an oscillatory behavior of the field inside the resonator and its evanescent outer components. The corresponding mode number is the axial mode number l which normally describes the number of intensity maxima along the axial direction. In some geometries, e.g., for spherical microstructures, the fundamental mode in the equatorial plane often refers to a condition $m = l$. In this special case, the deviation from this equality describes the number of intensity maxima in the axial direction.

For the azimuthal direction, Eq. (2.17) can directly be solved and the resulting solutions are simple complex exponentials of the form

$$\Phi(\phi) = e^{\pm im\phi} \tag{2.19}$$

which describe the periodicity of the circulating wave around the circumference of the resonator [48].

For the radial direction, Eq. (2.16) is the well-known Bessel equation. Their solutions are the Bessel functions of the first J_m and second Y_m kind. With them, the general solution of the equation can be found as [31]

$$U(\rho) = \begin{cases} A_m J_m(k_0 n_{eff} \rho) & \rho < R \\ A_m (J_m + iY_m)(k_0 \rho) = A_m H_m^{(1)}(k_0 \rho) & \rho > R \end{cases} \tag{2.20}$$

with R referring to the radius of the disk. The radial solutions inside the disk ($\rho < R$) are oscillatory Bessel functions, while the solutions outside the disk ($\rho > R$) are based on Hankel functions. These functions can be approximated by an exponential decay with decay constant $\alpha = k_0 \sqrt{n_{eff}^2 - n_0^2}$. This behavior reflects the evanescent component of the WGM outside the resonator and describes the approximate extension of the field in the environment.

When considering the different boundary conditions at the interface between resonator and environment, and by taking account for the continuity of the tangential fields across this interface, Eqs. (2.19) and (2.20) can be combined to give an analytic expression of the field distribution in the axial direction as

$$\Psi(\rho,\phi)=\begin{cases} A_m J_m(k_0 n_{eff}\rho)\ e^{\pm im\phi} & \rho < R \\ A_m \dfrac{J_m(k_0 n_{eff}R)}{H_m^{(1)}(k_0 R)} H_m^{(1)}(k_0\rho)\ e^{\pm im\phi} & \rho > R \end{cases}. \tag{2.21}$$

By using the continuity of the first order derivative, this equation can be solved. Often multiple individual resonances can be found for given azimuthal mode numbers *m*. The different solutions of the resulting transcendental equation are ordered by an additional parameter practically describing the number of intensity maxima along the radial direction. This order parameter is known as radial mode number *n*. In FIG. 2.5, the normalized energy density distributions of the three lowest of these radial mode orders are shown. The corresponding azimuthal dependence of the lowest order mode $n = 1$ is illustrated in FIG. 2.6. The third order parameter, the axial mode number *l*, is not explicitly shown here, but can be seen in FIG. 2.8d as result of a numerical simulation. When Maxwell's equations are applied to Eq. (2.21), the axial component also allows calculating the radial and azimuthal components of the corresponding magnetic and electric fields.

The various solutions of the Helmholtz equation are often represented by two symbols $\mathrm{TE}_{n,l}^{m}$ and $\mathrm{TM}_{n,l}^{m}$ with the integer mode numbers *m*, *n*, and *l* describing the number of intensity maxima of the transverse fields in azimuthal, radial, and axial direction.

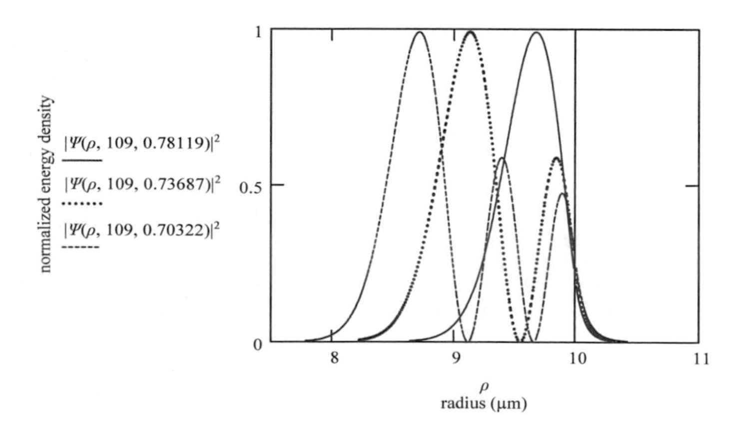

FIG. 2.5. Analytic solution for the normalized energy density distributions proportional to $|U(\rho, m, \lambda)|^2$ of the first three transverse magnetic H_z modes (TE) along the radial direction in a microdisk with radius $R = 10$ μm. For the azimuthal mode number $m = 109$, the corresponding resonance wavelength λ with radial mode numbers $n = \{1, 2, 3\}$ can be calculated by the continuity of the fields at the interface. The black line corresponds to the outer edge of the resonator. The evanescent fields start at the right side of that line.

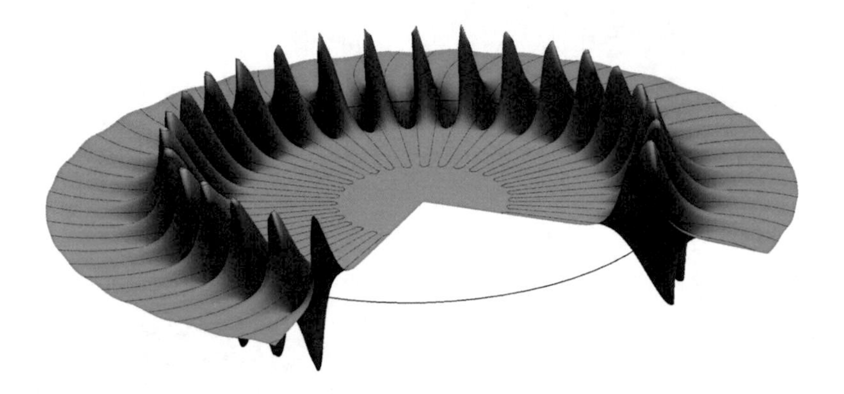

FIG. 2.6. Analytic solution for the axial electric E_z (TM) or magnetic H_z field (TE) in an ideal microdisk with azimuthal mode number $m = 32$ ($n = 1$). The circular black line represents the outer edge of the resonator.

2.3.2 Optical Modes of Spherical Microresonators

Another type of resonators which exhibits WGM resonances are the so-called spherical microresonators. These resonators are in between conventional macroscopic devices and more integrable devices produced by lithography. As they can be fabricated by simply melting the tips of standard silica glass fibers [49] or by heating up small pieces of falling glass [50], they were among the first objects which allowed an intensive investigation of optical WGMs. Spherical microresonators are very simple structures and thus they were extensively studied during the last few decades. For this type of resonators, full analytic solutions of Maxwell's equations can be found. Hence, such systems are very well understood. Although nowadays the chip-integrated systems, e.g., toroidal, disk, or wedge microresonators, become more and more a matter of interest, the theoretical background of spherical microresonators has still applications for these modern types of microresonators. In the case of novel hollow microspheres or microbubble resonators, the analytic solutions for microspheres are applicable and can be used to describe the modal structure of the devices.

Therefore, in the following section also the analytic solution of Maxwell's equations in spherical coordinates is presented. The mathematical techniques used to derive these solutions are widely equivalent to the techniques used to describe the optical modes in disk-type microresonators. A more detailed derivation of the presented solution for a spherical coordinate system can be found in [45].

After introducing polar coordinates $\mathbf{r} = (\rho, \varphi, \theta)$ to better describe the geometry of the resonator system, the Cartesian Helmholtz equation (Eq. (2.14)) can be written in their spherical form as

$$\left[\frac{\partial^2}{\partial\rho^2}+\frac{1}{\rho^2\sin\theta}\frac{\partial}{\partial\vartheta}\sin\theta\frac{\partial}{\partial\vartheta}+\frac{1}{\rho^2\sin^2\theta}\frac{\partial^2}{\partial\phi^2}+\frac{\partial^2}{\partial z^2}+k_0^2n(\boldsymbol{r})^2\right]\boldsymbol{\Psi}(\boldsymbol{r}) = 0\,. \quad (2.22)$$

Here, $\mathbf{\Psi}(\mathbf{r}) = \{\mathbf{E}(\mathbf{r}), \mathbf{H}(\mathbf{r})\}$ corresponds to the different field components at position $\mathbf{r}$, $k_0 = \omega/c$ is the wave number in free space, and $n(\mathbf{r}) = \sqrt{\varepsilon(\mathbf{r})\mu(\mathbf{r})}$ describes the local refractive index based on permittivity and permeability.

The solutions $\mathbf{\Psi}(\mathbf{r})$ of Eq. (2.22) are the well-known Debye potentials [51]. It has been shown that Eq. (2.22) can be fully solved by the method of variables separation when homogeneous dielectric spheres with polarization stable fields are assumed [45]. In this case, all solutions can be expressed by a single field component, i.e., E_φ or H_φ, in exact the same way as it was shown for dielectric microdisks. Hence, $\mathbf{\Psi}(\mathbf{r})$ can be separated into $\Psi(\mathbf{r}) = U(\rho)\Phi(\phi)\Theta(\theta)$. The group of the possible solutions again divides into two subgroups. These are labeled as transverse electric (TE) and transverse magnetic (TM) modes depending on which of the two fields is parallel to the surface of the sphere with radius R. The solutions of the three separated equations are given by [45]

$$\Phi(\phi) = e^{\pm im\phi} \quad (2.23)$$

$$U(\rho) = \begin{cases} A_l j_l(k_0 n_{eff}\rho) & \rho > R \\ A_l(j_l + iy_l)(k_0\rho) = A_l h_l^{(1)}(k_0\rho) & \rho > R \end{cases} \quad (2.24)$$

$$\Theta(\vartheta) = H_{l-m}(\sqrt{m}\theta)e^{-\frac{m}{2}\vartheta^2}\,. \quad (2.25)$$

These equations show a strong similarity to the solutions for disk-type resonators as examined in the last paragraph. They can again be ordered by three independent mode numbers. These are the radial mode number n, describing the number of intensity maxima in the radial direction, the azimuthal mode number m, describing the number of intensity maxima around the circumference of the sphere, and the polar mode number l, which indirectly corresponds to the number of intensity maxima along the polar direction. For the radial solution $U(\rho)$, the associated Bessel and Hankel functions are simply exchanged by their spherical equivalent, namely the spherical Bessel functions of the first and second kind, j_l and y_l, and the spherical Hankel function of the first kind $h_l^{(1)}$. In the azimuthal direction, the solution $\Phi(\phi)$ is identical with the solution for microdisks. However, the solution for the polar direction $\Theta(\theta)$ is coupled with the other two solutions by their corresponding mode numbers and includes a Hermite polynomial of the order $(l-m)$. Hereby, the fundamental mode among the different resonances can be found from the conditions $l = m$ and $n = 1$. For this specific set of parameters and some other values of $(l - m)$, the resulting energy density distributions are presented in FIG. 2.7.

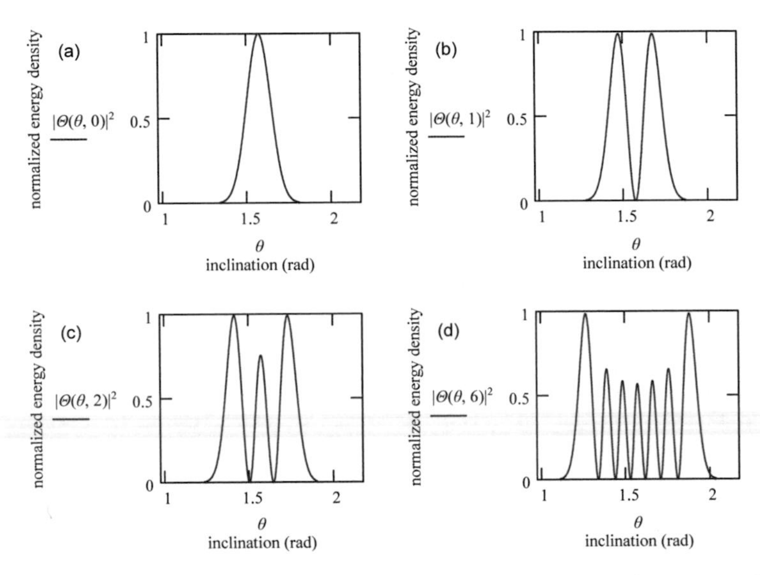

FIG. 2.7. Normalized energy density distributions of WGMs in a spherical microresonator. The analytic solutions for the different field components $\Theta(\vartheta, l - m)$ in polar direction are shown in dependence to the azimuthal mode number m and the polar mode number l. (a) $m = l = 109$ (fundamental mode). (b) $m = 109$, $l = 108$. (c) $m = 109$, $l = 107$. (d) $m = 109$, $l = 103$.

By solving the finally resulting characteristic equation, which connects the solutions of the interior and exterior fields at the surface of the sphere with $\rho = R$, the resonant wave numbers k are related to the mode numbers l and n. Directly from these relations, the corresponding resonance wavelengths can be calculated. The allowed mode numbers for m are integers ranging between $-|l| \leq m \leq +|l|$. Thereby, for modes with an azimuthal mode number $m > 0$, the direction of propagation is defined opposite to the direction of propagation for modes with $m < 0$.

In a perfect sphere, all modes that share a common value of $|m|$ are degenerated and thus they share the same resonance conditions and wavelengths. This degeneracy is lifted in oblate systems, i.e., in ellipsoidally deformed microspheres [43] or in spheres which are supported by a stem. The presence of a stem breaks the symmetry of the system and thus WGMs with unique mode numbers can be observed.

2.4 Numerical Simulation

For an exact analysis of the modal properties of WGMs in arbitrary microresonator structures, analytic methods are highly preferred. They allow simple insight into the underlying physical aspects and give direct access to the most important system parameters.

In the last section, the full analytic solutions of the Helmholtz equation for simple microdisks and spherical microspheres were presented. Also for some other types of microresonators exhibiting a circular symmetry, complete analytic solutions can be found [52]. These direct analytic approaches allow a simple analysis of the modal structure. Furthermore, it allows a direct calculation of the mode volumes, the FSRs, and the different resonance conditions in the resonators. The exact analysis of these properties with analytic methods normally requires hard approximations concerning the symmetry of the investigated system. In real disk-type or spherical microresonators, small deviations from a perfect symmetry often arise during manufacturing. While slightest deviations by compression and strain, e.g., from deforming the spherical shape of a microsphere into a more ellipsoidal shape, may still be considerable, more complex deviations based on isolated scattering centers or defects are hard to include into an analytic model. In disk-type microresonators, for example, slanted side walls caused by a non-isotropic etching behavior are challenging.

For toroidal microresonators, even under perfectly idealized conditions, a full analytic solution can not be derived [26]. This is due to the intrinsic non-separability of the Helmholtz equation in a toroidal coordinate system. By using a perturbative approach for the solution, a separation problem can be avoided, but the results are approximated. Furthermore, the method is limited to certain geometric aspect ratios between the major and minor diameters of the toroidal structure, and the perturbative calculation of the various mode distributions and intrinsic properties requires an iterative procedure. However, even with this technique, complex deformations of the resonator shape can not be fully considered.

An alternative to the presented rigorous analytic description of optical modes in WGM microresonators is given by using optical design tools. They perform software-based numerical simulations allowing a detailed analysis of highly complex resonator designs and structures even with strong deviations from a specific symmetry. For numerical simulations, a variety of different solving techniques are available. Depending on the specific problem, the most efficient solver approach can be selected. Most common are the finite element method (FEM), the finite-difference time-domain (FDTD) method, the beam propagation method (BPM) and the eigenmode expansion (EME) method. For analyzing the propagation properties in extended waveguide structures, the BPM and EME methods are preferred.

For the given numerical simulations presented with this thesis, mainly the EME-based commercial software suite Fimmwave including Fimmprop (Photon Design) was used. This package allows the full vectorial analysis of extended waveguides and tapered structures as well as a detailed analysis of the modes in arbitrary curved resonator designs. The EME method gives a direct access to the various coupling coefficients which connect the different eigenmodes of a complex photonic system. Furthermore, the transmission coefficients can also be calculated. For the field representation, the software uses the scattering matrix formalism. It allows a simple optimization of the coupling conditions between resonator and coupler. By this, the conditions can also be optimized for the interaction between specific modes, e.g., the fundamental modes in both systems, without the requirement of applying approximations to the underlying mode structures or the propagation constants. A more detailed description of the EME method can be found in Appendix A.

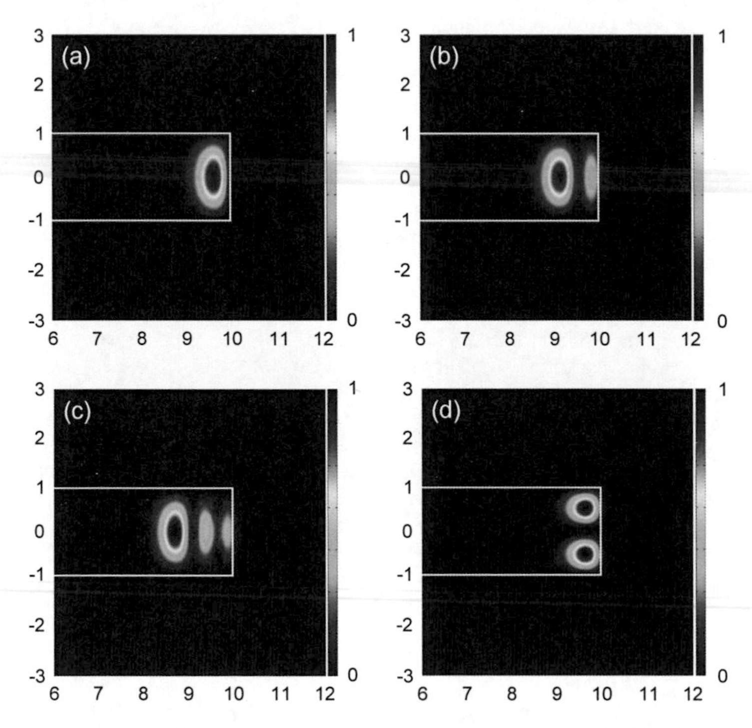

FIG. 2.8. Numerical simulation of the absolute magnetic energy density $|\mathbf{H}|^2$ in a silica disk microresonator with radius $R = 10$ µm. The results are in perfect agreement with the analytic solutions presented in Section 2.3.1. (a) $\mathrm{TE}_{1,1}^{109}$. (b) $\mathrm{TE}_{2,1}^{109}$. (c) $\mathrm{TE}_{3,1}^{109}$. (d) $\mathrm{TE}_{1,2}^{109}$.

For example, in FIGs. 2.8 – 2.10, the simulated energy density distributions for several types of WGM microresonators are presented. They are two-dimensional numerical simulations of the field distributions calculated with the EME module of Fimmwave. For calculating the eigenmodes of the systems, an FDTD-based solver was applied. The convergences of the different results were carefully proofed by probing the simulations with higher meshing parameters and by using another solver approach which is also provided by the simulation software (FEM).

In FIG. 2.8, numerical simulations of the absolute magnetic energy densities $|\mathbf{H}|^2$ for different modes in a disk-type silica microresonator are presented. The radius of the shown structure is $R = 10$ µm. The results are in perfect agreement with the analytic solutions given in Section 2.3.1. The model assumes a perfect symmetry for the solved structure.

The simulated absolute magnetic energy densities $|\mathbf{H}|^2$ for a spherical microresonator with $R = 10$ µm are shown in FIG. 2.9. In this case, the numbering of the modes is slightly different. The fundamental mode refers to the condition $\{m = l = 109, n = 1\}$ in reference to the analytical description of the structure by spherical Bessel functions and their underlying notations. Also in this special case, the symmetry of the structure is not disturbed and thus the presented results are in full agreement with the analytic approach introduced in Section 2.3.2.

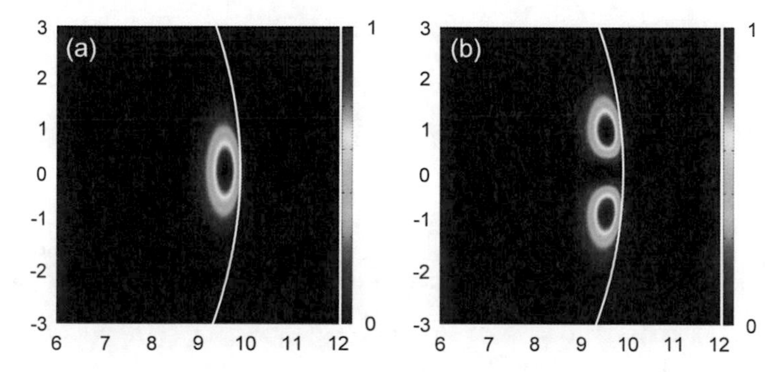

FIG. 2.9. Numerical simulation of the absolute magnetic energy density $|\mathbf{H}|^2$ in a silica sphere microresonator with radius $R = 10$ µm. (a) $\mathrm{TE}^{109}_{1,109}$. (b) $\mathrm{TE}^{109}_{1,108}$. The corresponding polar mode number is $l = (m - 1)$.

Other well-known types of WGM cavities are the so-called toroidal and wedge-type microresonators. Also for these specific designs the absolute magnetic energy density distribution of their respective fundamental modes was numerically simulated. The results are presented in FIG. 2.10. These resonator designs can not be fully analyzed with analytical methods and numerical techniques are obligatory to investigate the resulting mode structures. There are no numerical approximations required and even some specific shape deformations, e.g., caused by the processing techniques, can be implemented within the numerical model. The full numerical approach ensures realistic results with the highest possible accuracy.

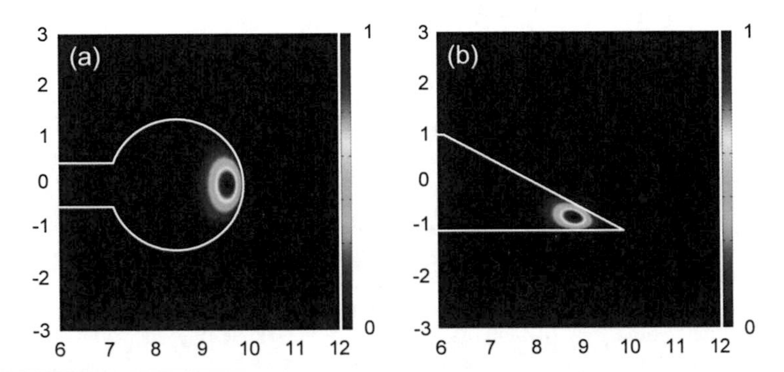

FIG. 2.10. Numerical simulation of the absolute magnetic energy density $|\mathbf{H}|^2$ in the resulting fundamental $TE_{1,1}^{109}$ modes for alternative types of microresonators both with radius $R = 10$ µm. (a) Toroidal microresonator. (b) Wedge resonator ($\alpha_{wedge} = 30°$).

2.5 Coupling to Whispering Gallery Modes

The excitation of WGMs in microresonators requires a possibility to efficiently couple photons into the cavity. Also the probing of different cavity resonances is based on interactions with external coupling devices. As the modes in a microresonator normally exhibit very high Q factors, the direct out-coupling due to scattering is very weak. With the reciprocity theorem, it can be shown that the probability for the inverse in-coupling process is also highly inefficient [53].

Another option for accessing WGMs is using photon tunneling due to the coupling of evanescent fields. If the coupling device and the resonator show at least a small spatial overlap in their field distributions, then they can interact with each other and photons can be tunneled with high efficiency. An important condition for an efficient coupling by photon tunneling is matching the individual propagation constants β of the resonator and the coupler. In other words, identical k vectors in both devices are required. This is to fulfill conservation of momentum for the tunneled photons and to ensure a coherent interaction along the extended coupling zone.

The technique of evanescent coupling to WGM microresonators is well-known and several coupler schemes have been developed, e.g., prism couplers [41], eroded single mode fibers [54], angle polished single mode fibers [55], or pedestal antiresonant reflecting waveguides [56]. For this thesis, mainly tapered optical fibers [57] were used. They allow great mechanical flexibility during their alignment. The micrograph inset in FIG. 2.11 shows a typical waist of such a fiber taper. All types of couplers are based on the same principal. A guided mode is exposed to the environment such that the evanescent field of that specific mode can interact with WGMs inside the resonator.

2.5.1 Transfer Matrix Formalism

In the following section, a very general description of the coupling process between the resonator and a waveguide coupler is presented. It is based on a scattering matrix formalism which is widely used also for the description of other common coupling phenomena [58]. With this technique, the fundamental spectral properties of a resonator can be derived by assuming a steady-state operation and matching the corresponding fields inside coupler and resonator.

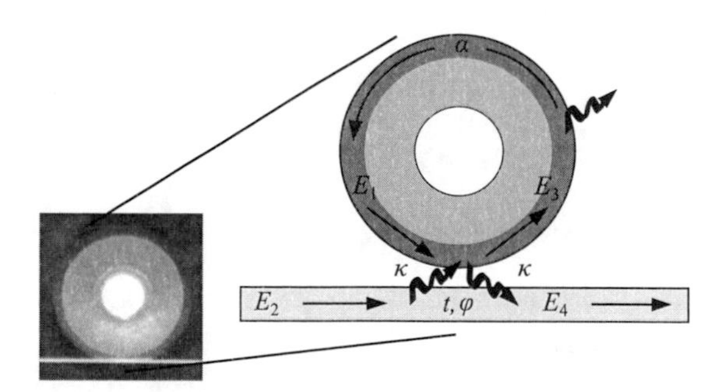

FIG. 2.11. Simple model of coupling between a circular cavity and a waveguide. The inset shows a micrograph of a typical micrometer-sized fiber taper coupled to a microdisk.

A major advantage of this formalism is the direct correspondence with the internal representation of numerical results derived by the EME-based numerical simulation tool. It uses a scattering matrix to describe the energy transfer through a boundary or a set of boundaries and allows mapping between the different modes of the input and output ports of an optical system. In case of coupling zones, the detailed energy transfer between the different modes of a waveguide and a WGM cavity can be traced along this path. For this, the software can be used to calculate the transmission, reflection, and loss coefficients for these modes. Also the internal cross-coupling coefficients can be calculated. The presented transfer matrix formalism is based on exactly the same coefficients and can directly be used for the further analysis. The derived numerical results can also be used for the investigation of more complex systems, e.g., coupled microresonators or photonic integrated circuits (PICs). This is highly advantageous over other analytic approaches for the description of the coupling process which are based on infinite sums [59] or rate equations [60]. A more detailed summary of the software-intern scattering matrix representation in the used EME-based simulation tool can be found in Appendix A.

The matrix formalism described in the following section is considering single-to-single mode interactions. The principal schematic of the formalism is depicted in FIG. 2.11. For a cavity resonance, the output field E_3 of the coupler is, after passing the resonator and acquiring a cavity dependent phase shift, feed back as the input field E_1. With the corresponding field amplitude vectors $\mathbf{a} = [E_1, E_2]^t$ and $\mathbf{b} = [E_3, E_4]^t$, this coupling condition can be written as

$$\mathbf{b} = \underline{\mathbf{S}} \cdot \mathbf{a}\,. \tag{2.26}$$

Here, **S** is the so-called scattering matrix which connects the different field amplitude vectors to each other. Due to power conservation and time-reversal, the general form of the scattering matrix is symmetrical and can be found as [32]

$$\underline{\mathbf{S}} = \begin{pmatrix} t_c & \kappa_c \\ \kappa_c & -t_c^* \dfrac{\kappa_c}{\kappa_c^*} \end{pmatrix} \tag{2.27}$$

with t_c and κ_c for the complex amplitude transmission and coupling coefficients. If no additional loss is considered, they fulfill the condition $\kappa_c^* t_c + t_c^* \kappa_c = 1$. The reference plane can be chosen such that $t_c = -t_c^* \kappa_c / \kappa_c^* = t \in \Re$. Defining also $\kappa \in \Re$, this results in $\kappa_c = \pm i\kappa$ and the scattering matrix can be written as

$$\underline{\mathbf{S}} = \begin{pmatrix} t & i\kappa \\ i\kappa & t \end{pmatrix} \text{ with } t^2 + \kappa^2 = 1\,. \tag{2.28}$$

Thus, for single mode waveguides, the coupling zone can be described by

$$\begin{pmatrix} E_3 \\ E_4 \end{pmatrix} = \begin{pmatrix} t & i\kappa \\ i\kappa & t \end{pmatrix} \cdot \begin{pmatrix} E_1 \\ E_2 \end{pmatrix}. \tag{2.29}$$

It can be assumed that the coupling and transmission coefficients are fully frequency independent because the effective coupling lengths are much smaller than the cavity circumference for most microresonators. Hence, this difference yields to a considerably broader bandwidth for the coupling than for the resonance of the cavity [61].

In the feedback path, the relation between the field amplitudes after one loop can be expressed by

$$E_1 = ae^{i\varphi}E_3, \tag{2.30}$$

where a is the single-pass amplitude transmission coefficient including all considered kinds of losses and φ is the acquired single-pass phase shift. The value of Eq. (2.30) is independent on adding or subtracting an integer multiple of 2π. Thus, for a resonance, the phase shift can be set to zero without loosing generality. For achieving constructive interference between the two field amplitudes E_1 and E_3, the phase shift has to be an integer multiple of 2π.

With the bulk material absorption coefficient α and the cavity roundtrip length L, an expression for the single-pass amplitude transmission coefficient a can be given as

$$a = e^{-\frac{\alpha}{2}L} = e^{-\frac{\alpha}{2}2\pi r_{eff}} = e^{-\alpha\pi r_{eff}}, \tag{2.31}$$

where r_{eff} is the effective radius of the corresponding cavity mode. The single-pass phase shift φ can be calculated from the cavity propagation constant β and the cavity roundtrip length L by

$$\varphi = \beta L = \frac{2\pi\ n_{eff}}{\lambda_0} 2\pi\ r_{eff} = 4\pi^2 \frac{n_{eff} r_{eff}}{\lambda_0} \tag{2.32}$$

with n_{eff} as the effective refractive index and λ_0 for the free-space wavelength of a specific cavity mode.

By solving the linear equation system consisting of Eqs. (2.29) – (2.32), the coupled resonator system can be analyzed in detail [62]. In the following paragraphs, the most important relations are presented to give an intuitive interpretation of the involved processes.

2.5.2 Intensity Buildup

A very interesting property of coupled microresonator systems is the fact that the optical intensity inside the cavity can highly exceed the initially injected intensity. In combination with the high energy densities, which are achievable due to the low mode volumes and high Q factors, this effect allows ultra-high intensities to circulate in the resonator even for pumping the waveguide with moderate input powers. This enables to observe intensity depended nonlinear effects, e.g., Raman lasing [63] or four-wave mixing [64], even in resonator materials which have nearly vanishing non-linear optical coefficients.

The cavity buildup factor can be calculated as the ratio between the circulating intensity of the field E_1 and the in-coupled waveguide intensity of the input field E_2. As it is based on constructive interference, an essential requirement for a cavity buildup is a sufficiently coherent light source. A constant phase relation is a basic requirement for observing interference effects and thus the coherence time of the input field has to be at least as long as the cavity photon lifetime. In case of a perfectly incoherent excitation, no intensity buildup can be observed and the intensities of the fields at the input and inside the cavity are equal. From Eqs. (2.29) – (2.32) and by squaring the modulus of the ratio between the fields E_1 and E_2, the cavity buildup factor can be defined as

$$K = \frac{I_1}{I_2} = \left|\frac{E_1}{E_2}\right|^2 = \left|\frac{i\kappa a e^{i\varphi}}{1 - ta e^{i\varphi}}\right|^2 . \tag{2.33}$$

By using Euler's formula, this can be further simplified to

$$K = \frac{(1-t^2)a^2}{1 - 2ta\cos\varphi + t^2 a^2} . \tag{2.34}$$

When the incident light is in resonance with the cavity, then the phase angle φ can be neglected. For weak coupling conditions ($\kappa \ll 1$) and by suppressing any single-pass losses ($a = 1$), this gives

$$K = \frac{1+t}{1-t} \approx \frac{4}{\kappa^2} . \tag{2.35}$$

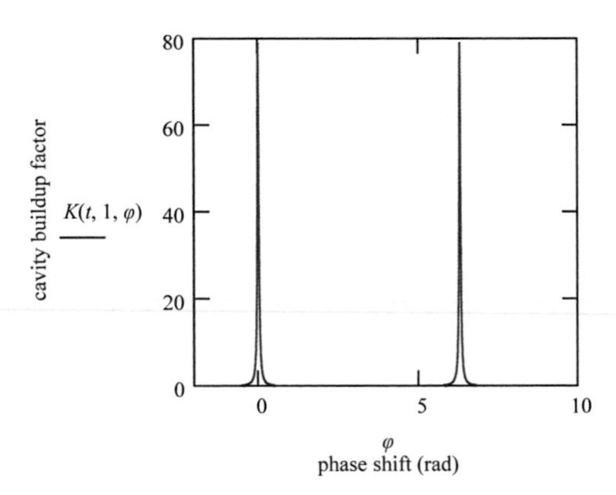

FIG. 2.12. Cavity buildup factor $K(t, a, \varphi)$ versus phase shift φ. Lossless single pass amplitude transmission $a = 1$ and a coupling coefficient $t = 0.975$ were assumed.

At couplings of 5 % ($\kappa^2 = 0.05$), the steady-state resonator intensity can reach 80 times the intensity of the input field. In FIG. 2.12, the dependence of the cavity buildup factor on the phase shift is illustrated. It can be seen that even without an explicit cavity loss factor the coupler itself already induces coupling losses which broadens the resonance linewidth. This linewidth directly corresponds to an external Q factor as introduced in the last section. As this kind of coupling loss is an intrinsic property of the couplers, weaker coupling conditions lead to higher Q factors and thus increase the maximum cavity buildup factor.

2.5.3 Coupled Finesse

The relation between the fields E_1 and E_2 allows calculating the finesse of the coupled system. As finesse is defined as the ratio between the FSR and the spectral linewidth of a resonance, the previous analysis of the cavity buildup factor can be used. A resonance linewidth is often represented by the full width at half maximum (FWHM). This value can directly be calculated from Eq. (2.33) by

$$FWHM = 2\arccos\left(\frac{2ta}{1+t^2a^2}\right). \tag{2.36}$$

By assuming weak coupling conditions ($\kappa << 1$) and small single-pass losses ($ta \approx 1$), the finesse can be found as

$$F = \frac{2\pi}{FWHM} \approx \frac{\pi}{1-ta} \approx \frac{2\pi}{\kappa^2 + \alpha L}. \tag{2.37}$$

The finesse is 2π times the number of roundtrips N inside the cavity and thus it can be concluded that $N \approx 1/(\kappa^2 + \alpha L)$. Hence, the finesse is directly related to the photon lifetime and the effective path length inside a resonator.

2.5.4 Transmission Spectrum

For analyzing the different WGMs of a microresonator, the transmission spectrum of the involved coupler is also an important property. This spectrum can be calculated from Eqs. (2.29) – (2.32) by squaring the modulus of the ratio between the input and output fields E_4 and E_2 as

$$T = \frac{I_4}{I_2} = \left|\frac{E_4}{E_2}\right|^2 = \left|e^{i(\varphi+\psi)}\frac{a - te^{-i\varphi}}{1 - tae^{i\varphi}}\right|^2 = \frac{t^2 + a^2 - 2at\cos(\varphi)}{1 + a^2t^2 - 2at\cos(\varphi)} \tag{2.38}$$

with an additional phase shift ψ acquired from passing the coupling zone.

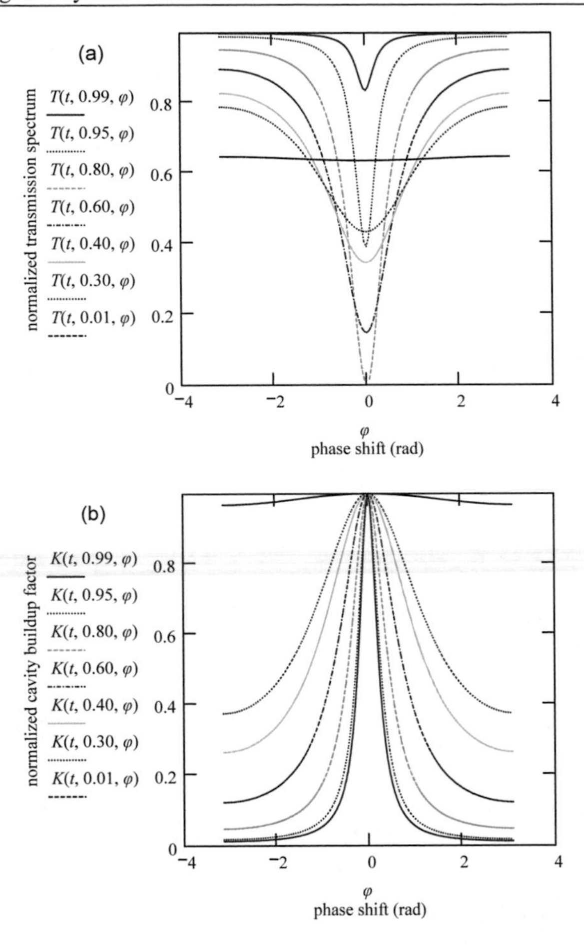

FIG. 2.13. (a) Transmission spectrum $T(t, a, \varphi)$ versus the acquired phase shift φ for different loss coefficients a. (b) Corresponding intensity spectrum of the normalized cavity buildup factor $K(t, a, \varphi)$. The amplitude transmission coefficient is $t = 0.8$.

If the single-pass loss of a cavity is negligible ($a = 0$), the transmission through the coupler is independent on a phase shift inside the resonator. In this case, the coupler transmission is unity for all frequencies and the device is referred to as an "all-pass" filter. This result can be understood by remembering the steady-state approach of the model. After initially loading the resonator, the intensity is stabilized corresponding to the specific cavity buildup factor of the system. Due to energy conservation, the optical powers entering and leaving the system have to be equal and thus unity transmission occurs for on- and off-resonance frequencies.

In real systems, internal attenuation processes can not be fully avoided and must be considered. In FIG. 2.13, sharp dips at the position of a resonance are noticeable in the transmission spectrum when resonator losses are fully taken into account. The dips correspond to different coupling conditions and show a clear dependency on the ratio between the transmission and the loss coefficient. The transmission spectrum can be separated into three different coupling regimes:

> If the given transmission coefficient t and loss coefficient a are equal, then the transmission through the output port of the coupler is zero and **critical** coupling is achieved. In this case, the directly transmitted field components from the input port cancel out with the components coupled back from the resonator. The whole optical power entering the waveguide is coupled into the cavity. The circulating intensity inside the resonator reaches maximum and the dip reaches the zero level. As the two coefficients are directly related to the intrinsic and extrinsic Q factors, this condition can be interpreted as $Q_0^{-1} = Q_{ext}^{-1}$.
>
> If the loss coefficient is higher than the transmission coefficient, i.e., $Q_0^{-1} > Q_{ext}^{-1}$, the corresponding regime is called **under-coupled**. The resonator is only weakly coupled and the intrinsic losses dominate. The out-coupled field of the resonator is small compared to the directly transmitted field of the waveguide. For that regime, a high off-resonance coupler transmission and a very narrow resonance linewidth are characteristic. The dip does not reach the zero level.
>
> The same stands for the so-called **over-coupled** working regime. In this case, the actual loss coefficient is lower than the transmission coefficient, i.e., $Q_0^{-1} < Q_{ext}^{-1}$. The out-coupled field from the resonator is higher than the transmitted field of the waveguide. The intensity far from the resonance is highly suppressed and the linewidth of the dip is strongly broadened in comparison to the under-coupled and critical regimes.

For working with microresonators, the critical coupling regime is often preferred due to the achievable high intensities. As the loss of a resonator, expressed by its single-pass amplitude transmission coefficient a, is more or less an intrinsic property of the system, the transmission coefficient t is the only free parameter for a control of the coupling conditions. It can be often changed by simply varying the distance between waveguide and resonator. Due to this variation, also the modal coupling between cavity and waveguide fields is modified and critical coupling can be established. However, by approaching the waveguide to the resonator, additional coupling loss may appear. This effect is not considered here, but can be implemented within the model by introducing distance dependent transmission and coupling coefficients.

2.5.5 Coupling Conditions

For determining the optimum coupling conditions between the resonator and a coupled waveguide, e.g., an optical fiber or an integrated structure, two different parameters have to be considered. The first one is the field overlap between the waveguide mode in the coupler and the WGM inside the cavity; the second one is the degree of phase matching between these two modes. In a high-Q cavity, coupling is weak and the respective coupling coefficients are small. In this case, the modes in waveguide and cavity can be seen as unperturbed fields and a coupled mode approach is applicable. Therefore, the total amplitude coupling coefficient κ can be calculated from these fields by using an overlap integral. With the propagation constants of the modes, this integral can be defined as [65]

$$\kappa \propto (n_c^2 - n_0^2) \iiint_V \left(\mathbf{E}_w^t \cdot \mathbf{E}_c^{t*} + E_w^z \cdot E_c^{z*} \right) e^{-i(\beta_w - \beta_c^z)z} dV \,. \tag{2.39}$$

Here, n_c and n_0 are the refractive indices of the cavity and for air, $\mathbf{E}_{w,c}^t$ and E_w^z are the transverse and longitudinal field components in the cavity and in the waveguide, and $\beta_{c,w}$ are the corresponding propagation constants. The highest coupling coefficients are observed when the different fields are strongly interacting, i.e., when they have a large evanescent field overlap, and their propagation constants are matched to each other. As shown in FIG. 2.14, a phase matching condition $\Delta\beta = \beta_w - \beta_c \equiv 0$ is required to ensure maximum overlap along the direction of propagation.

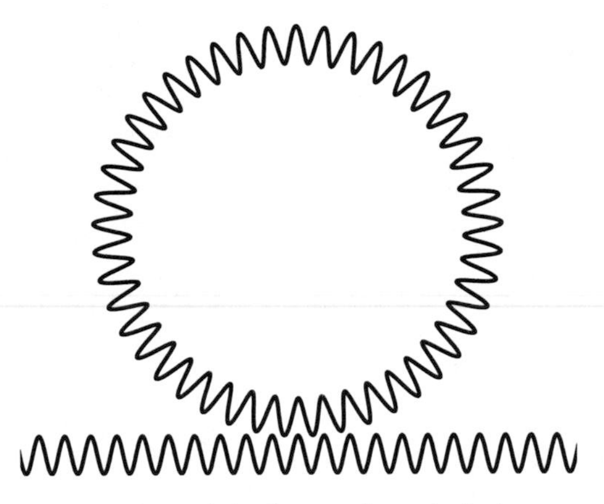

FIG. 2.14. Illustration of the phase relation between the modes in the resonator and a coupled waveguide (from [66]).

In FIG. 2.14, it can also be seen that coupling to a curved structure forces the phase correlation to become spatially depended. A mismatch occurs everywhere except for the point of closest proximity. Hence, the propagation constant of a cavity mode must be projected to the straight waveguide and an effective propagation constant has to be considered. For the point of closest proximity, the resulting modified phase matching condition can be given as [26]

$$\beta_w = \beta_{c,eff} \approx \left[1 - \frac{r_w + d_{gap}}{2r_c}\right] \beta_c \tag{2.40}$$

with r_w and r_c for the radii of waveguide and cavity, and d_{gap} as their mutual spacing. Therefore, the propagation constant of a waveguide, which is used to couple light into a resonator, is required to be fractionally thinner than the value given by the simple linear phase matching criterion.

2.5.6 Multimode Coupling

The scattering matrix formalism as presented in the last section is based on a simple single-mode approximation. The waveguide as well as the resonator are considered to support only one propagational mode. The two fields can only interact with each other or loose energy due to energy dissipation processes.

In real systems, normally a whole set of different modes is interacting with each other and a single-mode approximation may not be sufficient anymore. A standard fiber taper often supports multiple optical modes. When using such a device for out-coupling the WGMs of a resonator, all the taper modes are excited with different coupling strength. Because of the given phase matching condition often only the lowest order modes are prominent. The other modes are just acting as additional sources of loss. The strength of such parasitic coupling losses can be full quantified by a special parameter called "ideality" [67]. It is defined as the ratio of power coupled to a desired coupler mode divided by the overall-power coupled or lost to all other modes. For microresonators coupled to fiber tapers, high ideality values in excess of 99.97 % could be reached under appropriate conditions [67]. In these cases, mainly the two fundamental modes are interacting with each other and single-mode operation can still be considered.

For other systems exhibiting a lower value of ideality, the interaction of multiple modes has to be considered. The presented scattering matrix formalism is fully applicable also for such multimode systems. It can be widely adapted by simply building vectors out of the scalars which represent the different field amplitudes of the various modes inside the structures. By also fully considering the individual cross-coupling and amplitude transmission coefficients between these specific modes, the scattering matrix becomes a supermatrix and the Eqs. (2.29) and (2.30) can be replaced by an equivalent vector representation given by [68]

$$\begin{pmatrix} \mathbf{E}_3 \\ \mathbf{E}_4 \end{pmatrix} = \begin{pmatrix} \underline{\mathbf{S}_{11}} & \underline{\mathbf{S}_{12}} \\ \underline{\mathbf{S}_{21}} & \underline{\mathbf{S}_{22}} \end{pmatrix} \begin{pmatrix} \mathbf{E}_1 \\ \mathbf{E}_2 \end{pmatrix} \tag{2.41}$$

$$\begin{bmatrix} E_1 \\ \vdots \\ E_n \end{bmatrix}_2 = \begin{bmatrix} e^{-\frac{1}{2}\alpha_1 L_1} e^{i\beta_1 L_1} & 0 & 0 \\ 0 & \ddots & 0 \\ 0 & 0 & e^{-\frac{1}{2}\alpha_n L_1} e^{i\beta_1 L_n} \end{bmatrix} \cdot \begin{bmatrix} E_1 \\ \vdots \\ E_n \end{bmatrix}_4 . \tag{2.42}$$

Here, the $\underline{\mathbf{S}_{mn}}$ are the transfer matrices for mapping between the different modal fields of the input and output ports. E_i is the field amplitude of the i-th mode inside such a multimode waveguide port.

This fully vector-based formalism also allows for considering the interactions between different modes of a system and thus the quasi-analytic description of real systems becomes possible with this method. However, to get full access to all the required coefficients, other techniques have to be applied. An analytic calculation is possible by considering the field overlaps between different modes. Another option is using numerical simulations. With a photonic design tool, all important modal properties, e.g., the transmission, reflection, cross-correlation, and loss coefficients, can directly be calculated. The used EME-based software suite from Photon Design (Fimmwave) is perfectly suited for such simulations as the results are already presented in the required scattering matrix form (see Appendix A). Thus, arbitrary resonator-coupler designs with multiple optical modes can be numerically investigated and further analyzed by the presented multimode scattering matrix formalism.

[1] A. Siegman, "Laser beams and resonators: beyond the 1960s," *J. Sel. Top. Quantum Electron.*, vol. 6, no. 6, pp. 1389–1399, 2000.

[2] G. Indebetouw, "Tunable spatial filtering with a Fabry-Perot etalon," *Appl. Opt.*, vol. 19, no. 5, pp. 761–764, 1980.

[3] I. Ozdur, D. Mandridis, M. U. Piracha, M. Akbulut, N. Hoghooghi, and P. J. Delfyett, "Optical frequency stability measurement using an etalon-based optoelectronic oscillator," *Photonics Technol. Lett.*, vol. 23, no. 4, pp. 263–265, 2011.

[4] T. Ikegami and K. Kubodera, "Nonlinear optical devices for switching applications," in *IEEE International Conference on Communications (ICC)*, 1990, vol. 3, pp. 1152–1156.

[5] M. Ménard and A. Kirk, "Broadband integrated Fabry-Perot electro-optic switch," in *International Conference on Photonics in Switching (PS)*, 2008, pp. 1–2.

[6] M. J. Strain and M. Sorel, "Integrated III-V Bragg gratings for arbitrary control over chirp and coupling coefficient," *Photonics Technol. Lett.*, vol. 20, no. 22, pp. 1863–1865, 2008.

[7] A. Wang, Y. Wang, and H. He, "Enhancing the bandwidth of the optical chaotic signal generated by a semiconductor laser with optical feedback," *Photonics Technol. Lett.*, vol. 20, no. 19, pp. 1633–1635, 2008.

[8] B. D. Timotijevic, F. Y. Gardes, W. R. Headley, G. T. Reed, M. J. Paniccia, O. Cohen, D. Hak, and G. Z. Masanovic, "Multi-stage racetrack resonator filters in silicon-on-insulator," *J. Opt. A Pure Appl. Opt.*, vol. 8, no. 7, pp. 473–476, 2006.

[9] S. Khuntaweetep, S. Somkuarnpanit, and K. Sae-Tang, "Determination of bandwidth and free spectral range for the silicon based ring resonators and racetrack microcavity resonators," in *Asia-Pacific Conference on Circuits and Systems (APCCAS)*, 2004, vol. 1, pp. 497–499.

[10] Lord Rayleigh, "The problem of the whispering gallery," *Philos. Mag.*, vol. 20, no. 120, pp. 1001–1004, 1910.

[11] P. D. Dapkus, S. J. Choi, S. J. Choi, K. Djordjev, T. Sadagopan, and D. Tishinin, "Microresonator devices for DWDM systems," in *Optical Fiber Communication Conference and Exhibit (OFC)*, 2001, vol. 3, p. WK1.

[12] B. E. Little, J. S. Foresi, G. Steinmeyer, E. R. Thoen, S. T. Chu, H. A. Haus, E. P. Ippen, L. C. Kimerling, and W. Greene, "Ultra-compact Si-SiO_2 microring resonator optical channel dropping filters," *Photonics Technol. Lett.*, vol. 10, no. 4, pp. 549–551, 1998.

[13] P. Dong, S. F. Preble, and M. Lipson, "All-optical compact silicon comb switch," *Opt. Express*, vol. 15, no. 15, pp. 9600–9605, 2007.

[14] A. Canciamilla, M. Torregiani, C. Ferrari, F. Morichetti, R. M. De La Rue, A. Samarelli, A. Sorel, and A. Melloni, "Silicon coupled-ring resonator structures for slow light applications: potential, impairments and ultimate limits," *J. Opt.*, vol. 12, no. 10, p. 104008, 2010.

[15] M. Gregor, C. Pyrlik, R. Henze, A. Wicht, A. Peters, and O. Benson, "An alignment-free fiber-coupled microsphere resonator for gas sensing applications," *Appl. Phys. Lett.*, vol. 96, no. 23, p. 231102, 2010.

[16] J. T. Robinson, L. Chen, and M. Lipson, "On-chip gas detection in silicon optical microcavities," *Opt. Express*, vol. 16, no. 6, pp. 4296–4301, 2008.

[17] M. Hauser, "Mikroresonatoren aus Glas und Polymeren als optische Flüstergalerien," Karlsruher Institut für Technologie, 2011.

[18] A. M. Armani, R. P. Kulkarni, S. E. Fraser, R. C. Flagan, and K. J. Vahala, "Label-free, single-molecule detection with optical microcavities," *Science (80-.).*, vol. 317, no. 5839, pp. 783–787, 2007.

[19] F. Vollmer, D. Braun, A. Liebchaber, M. Khoshsima, I. Teraoka, and S. Arnold, "Protein detection by optical shift of a resonant microcavity," *Appl. Phys. Lett.*, vol. 80, no. 21, pp. 4057–4059, 2002.

[20] M. Gregor, "Fiber Taper-Coupled Microresonators for Applications in Sensing and Quantum Optics," Humboldt-Universität zu Berlin, 2011.

[21] I. H. Agha, J. E. Sharping, M. A. Foster, and A. L. Gaeta, "Optimal sizes of silica microspheres for linear and nonlinear optical interactions," *Appl. Phys. B*, vol. 83, no. 2, pp. 303–309, 2006.

[22] P. Del'Haye, A. Schliesser, and O. Arcizet, "Optical frequency comb generation from a monolithic microresonator," *Nature*, vol. 450, no. 7173, pp. 1214–1217, 2007.

[23] T. J. Kippenberg and K. J. Vahala, "Cavity optomechanics: back-action at the mesoscale," *Science (80-.).*, vol. 321, no. 5893, pp. 1172–1176, 2008.

[24] W. von Klitzing, R. Long, V. S. Ilchenko, J. Hare, and V. Lefèvre-Seguin, "Tunable whispering gallery modes for spectroscopy and CQED experiments," *New J. Phys.*, vol. 3, no. 14, pp. 1–14, 2001.

[25] A. B. Matsko and V. S. Ilchenko, "Optical resonators with whispering-gallery modes—part I: basics," *J. Sel. Top. Quantum Electron.*, vol. 12, no. 1, pp. 3–14, 2006.

[26] B. Min, L. Yang, and K. J. Vahala, "Perturbative analytic theory of an ultrahigh-Q toroidal microcavity," *Phys. Rev. A*, vol. 76, no. 1, p. 013823, 2007.

[27] J. Gao, P. Heider, C. J. Chen, X. Yang, C. A. Husko, and C. Wei Wong, "Observations of interior whispering gallery modes in asymmetric optical resonators with rational caustics," *Appl. Phys. Lett.*, vol. 91, no. 18, p. 181101, 2007.

[28] A. N. Oraevsky, “Whispering-gallery waves,” *Quantum Electron.*, vol. 32, no. 5, pp. 377–400, 2002.

[29] J. D. Jackson, *Classical Electrodynamics*, 3rd ed. New York: Wiley, 1998.

[30] B. E. A. Saleh and M. C. Teich, “Fundamentals of Photonics,” in *Wiley Series in Pure and Applied Optics*, 2nd ed., vol. 32, Hoboken: Wiley, 2007.

[31] M. Borselli, T. J. Johnson, and O. Painter, “Beyond the Rayleigh scattering limit in high-Q silicon microdisks: theory and experiment,” *Opt. Express*, vol. 13, no. 5, pp. 1515–1530, 2005.

[32] H. A. Haus, *Waves and fields in optoelectronics*. Englewood Cliffs: Prentice-Hall, 1984.

[33] M. L. Gorodetsky, A. A. Savchenkov, and V. S. Ilchenko, “Ultimate Q of optical microsphere resonators,” *Opt. Lett.*, vol. 21, no. 7, pp. 453–455, 1996.

[34] H. Lee, T. Chen, J. Li, K. Y. Yang, S. Jeon, O. Painter, and K. J. Vahala, “Chemically etched ultrahigh-Q wedge-resonator on a silicon chip,” *Nat. Photonics*, vol. 6, no. 6, pp. 369–373, 2012.

[35] R. Henze, C. Pyrlik, A. Thies, J. M. Ward, A. Wicht, and O. Benson, “Fine-tuning of whispering gallery modes in on-chip silica microdisk resonators within a full spectral range,” *Appl. Phys. Lett.*, vol. 102, no. 4, p. 041104, 2013.

[36] D. K. Armani, T. J. Kippenberg, S. M. Spillane, and K. J. Vahala, “Ultra-high-Q toroid microcavity on a chip,” *Nature*, vol. 421, no. 6926, pp. 925–928, 2003.

[37] M. Borselli, K. Srinivasan, P. E. Barclay, and O. Painter, “Rayleigh scattering, mode coupling, and optical loss in silicon microdisks,” *Appl. Phys. Lett.*, vol. 85, no. 17, pp. 3693–3695, 2004.

[38] M. L. Gorodetsky, A. D. Pryamikov, and V. S. Ilchenko, “Rayleigh scattering in high-Q microspheres,” *J. Opt. Soc. Am. B*, vol. 17, no. 6, pp. 1051–1057, 2000.

[39] R. Kitamura, L. Pilon, and M. Jonasz, “Optical constants of silica glass from extreme ultraviolet to far infrared at near room temperature,” *Appl. Opt.*, vol. 46, no. 33, pp. 8118–8133, 2007.

[40] A. M. Armani, A. Srinivasan, and K. J. Vahala, “Soft lithographic fabrication of high Q polymer microcavity arrays,” *Nano Lett.*, vol. 7, no. 6, pp. 1823–1826, 2007.

[41] V. B. Braginsky, M. L. Gorodetsky, and V. S. Ilchenko, “Quality-factor and nonlinear properties of optical whispering-gallery modes,” *Phys. Lett. A*, vol. 137, no. 7–8, pp. 393–397, 1989.

[42] G. Agrawal, “Nonlinear fiber optics,” *Lect. Notes Phys.*, vol. 542, pp. 195–211, 2000.

[43] S. Schiller and R. L. Byer, "High-resolution spectroscopy of whispering gallery modes in large dielectric spheres," *Opt. Lett.*, vol. 16, no. 15, pp. 1138–1140, 1991.

[44] G. Righini, Y. Dumeige, P. Féron, M. Ferrari, G. Nunzi Conti, D. Ristic, and S. Soria, "Whispering gallery mode microresonators: fundamentals and applications," *Riv. del nuovo Cim.*, vol. 34, no. 7, pp. 435–486, 2011.

[45] B. E. Little, J.-P. Laine, and H. A. Haus, "Analytic Theory of Coupling from Tapered Fibers and Half-Blocks into Microsphere Resonators," *J. Light. Technol.*, vol. 17, no. 4, pp. 704–715, 1999.

[46] M. Born and E. Wolf, *Principles of Optics*. London: Pergamon, 1980.

[47] A. Morand, K. Phan-Huy, Y. Desieres, and P. Benech, "Analytical study of the microdisk's resonant modes coupling with a waveguide based on the perturbation theory," *J. Light. Technol.*, vol. 22, no. 3, pp. 827–832, 2004.

[48] B. E. Little and S. T. Chu, "Estimating surface-roughness loss and output coupling in microdisk resonators," *Opt. Lett.*, vol. 21, no. 17, pp. 1390–1392, 1996.

[49] D. W. Vernooy, V. S. Ilchenko, H. Mabuchi, E. W. Streed, and H. J. Kimble, "High-Q measurements of fused-silica microspheres in the near infrared," *Opt. Lett.*, vol. 23, no. 4, pp. 247–249, 1998.

[50] J. M. Ward and O. Benson, "WGM microresonators: sensing, lasing and fundamental optics with microspheres," *Laser Photon. Rev.*, vol. 570, no. 4, pp. 553–570, 2011.

[51] P. Debye, "Der Lichtdruck auf Kugeln von beliebigem Material," *Ann. Phys.*, vol. 335, no. 11, pp. 57–136, 1909.

[52] M. Sumetsky, Y. Dulashko, J. M. Fini, A. Hale, and D. J. DiGiovanni, "The microfiber loop resonator: theory, experiment, and application," *J. Light. Technol.*, vol. 24, no. 1, pp. 242–250, 2006.

[53] S. Lacey, H. Wang, D. H. Foster, and J. U. Nöckel, "Directional tunneling escape from nearly spherical optical resonators," *Phys. Rev. Lett.*, vol. 91, no. 3, p. 033902, 2003.

[54] N. Dubreuil, J. C. Knight, D. K. Leventhal, V. Sandoghdar, J. Hare, and V. Lefèvre, "Eroded monomode optical fiber for whispering-gallery mode excitation in fused-silica microspheres," *Opt. Lett.*, vol. 20, no. 8, pp. 813–815, 1995.

[55] V. S. Ilchenko, X. S. Yao, and L. Maleki, "Pigtailing the high-Q microsphere cavity: a simple fiber coupler for optical whispering-gallery modes," *Opt. Lett.*, vol. 24, no. 11, pp. 723–725, 1999.

[56] B. E. Little, J.-P. Laine, D. R. Lim, H. A. Haus, L. C. Kimerling, and S. T. Chu, "Pedestal antiresonant reflecting waveguides for robust coupling to microsphere resonators and for microphotonic circuits," *Opt. Lett.*, vol. 25, no. 1, pp. 73–75, 2000.

[57] J. C. Knight, G. Cheung, F. Jacques, and T. A. Birks, "Phase-matched excitation of whispering-gallery-mode resonances by a fiber taper," *Opt. Lett.*, vol. 22, no. 15, pp. 1129–1131, 1997.

[58] A. F. Sadreev, E. N. Bulgakov, and I. Rotter, "S-matrix formalism of transmission through two quantum billiards coupled by a waveguide," *J. Phys. A. Math. Gen.*, vol. 38, no. 40, pp. 10647–10661, 2005.

[59] M. Cai, O. Painter, and K. J. Vahala, "Observation of critical coupling in a fiber taper to a silica-microsphere whispering-gallery mode system," *Phys. Rev. Lett.*, vol. 85, no. 1, pp. 74–77, 2000.

[60] K. J. Vahala, Ed., "Optical Microcavities," in *Advanced Series in Applied Physics*, vol. 5, Singapore: World Scientic, 2004.

[61] C.-S. Ma, Y.-Z. Xu, X. Yan, Z.-Q. Qin, and X.-Y. Wang, "Optimization and analysis of series-coupled microring resonator arrays," *Opt. Commun.*, vol. 262, pp. 41–46, 2006.

[62] J. Heebner, R. Grover, and T. Ibrahim, Eds., "Optical Microresonators," in *Springer Series in Optical Sciences*, vol. 138, London: Springer, 2008.

[63] M. Jouravlev, D. R. Mason, and K. S. Kim, "Ultralow Raman lasing threshold and enhanced gain of whispering gallery modes in silica microspheres," *Phys. Rev. A*, vol. 85, no. 1, p. 013825, 2012.

[64] I. H. Agha, Y. Okawachi, and A. L. Gaeta, "Theoretical and experimental investigation of broadband cascaded four-wave mixing in high-Q microspheres," *Opt. Express*, vol. 17, no. 18, pp. 16209–16215, 2009.

[65] B. E. Little, S. T. Chu, H. A. Haus, J. Foresi, and J.-P. Laine, "Microring resonator channel dropping filters," *J. Light. Technol.*, vol. 15, pp. 998–1005, 1997.

[66] S. M. Spillane, "Fiber-coupled ultra-high-Q microresonators for nonlinear and quantum optics," California Institute of Technology, 2004.

[67] S. M. Spillane, T. J. Kippenberg, O. Painter, and K. J. Vahala, "Ideality in a fiber-taper-coupled microresonator system for application to cavity quantum electrodynamics," *Phys. Rev. Lett.*, vol. 91, no. 4, p. 043902, 2003.

[68] D. F. G. Gallagher and T. P. Felici, "Eigenmode expansion methods for simulation of optical propagation in photonics: pros and cons," in *Proceedings of SPIE*, 2003, vol. 4987, pp. 69–82.

Chapter 3

Whispering Gallery Mode Microbubble Resonators

3.1 Introduction

A relatively new type of WGM resonator is the so-called microbubble resonator. In principle, this is essentially the hollow version of a conventional spherical or bottleneck microresonator [1], [2]. Microbubbles are often produced by heating and pressurizing standard glass capillaries [3]. Thereby, the capillary openings allow a direct access to the inner volume of the cavity via the supporting stem. This allows modifications on the external and internal environment of the microresonators. The access may be used for filling the inside of a microbubble with specific liquids or gases. By assuming a sufficiently thin wall thickness of the microbubble, the circulating whispering gallery modes (WGMs) can evanescently couple to the environments on both sides of the resonator wall. This also allows, for instance, an effective separation of an external coupling device, e.g., a tapered optical fiber or a prism, from the variable conditions inside the cavity. Therefore, this type of resonator is a very promising candidate for applications in microfluidic sensing. Otherwise, liquid or gas at the outside of the resonator can strongly disturb the necessarily stable fiber coupler alignment [4]. For instance, within liquid-core optical ring-resonators (LCORRs), a small change in the refractive index of an interior media is detected by the inner evanescent field of an externally excited WGM [5].

After filling such a microbubble, it is also possible to modify the dimensions and optical properties of the resonator by pressurizing the inner medium. This also affects the WGMs inside the cavity and allows for simple resonance frequency tuning via an internal pressure change. By design, microbubbles are ideally suited for applications requiring narrow resonance linewidths in conjunction with large possible tuning ranges.

A pressure tuning method is very appealing for this type of microresonator. Especially for cryogenic applications, tuning methods are required that do not induce an additional thermal load to the system. The pressure tuning mechanism of microbubbles can be used as an effective alternative to other mechanical approaches for working under cryogenic conditions like strain tuning [6] or dimensional tuning by controlled nitrogen layer deposition [7].

In this chapter, this new tuning mechanism for WGM microbubble resonators is presented. After a brief introduction to the special properties of such resonator systems, their manufacturing process is explained in detail. For the characterization of the produced microbubbles, a model for estimating the cavity wall thickness from directly measurable geometric pre-production parameters is given. After modeling the pressure tuning of microbubbles, these theoretic results are compared to experimental room temperature tuning data. For extending this experimental analysis into the cryogenic temperature range, a home-build cryostat system was developed and is here explained in detail. With this system, the temperature dependent resonance shift by cooling the resonator from room temperature to liquid nitrogen (LN) temperature could also be examined.

The presented experimental works for estimating the intrinsic tuning properties of WGMs in microbubbles were performed in terms of possible cryogenic applications in quantum optics and resulted in two different publications [8], [9].

3.2 Optical Modes

The theoretical description of WGMs in microbubble resonators is very similar to the description of the optical modes in microspheres as it can be found in Chapter 2. In contrast to the fully homogeneous microsphere, for a microbubble also the dielectric properties of material inside the resonator have additionally to be taken into account. The parameters of the model and a general schematic of the basic geometry are shown in FIG. 3.1.

As for ordinary microbubbles, a separation approach can be used for modeling the electric field distribution $E(r,\vartheta,\varphi)$ inside a microbubble. With Y_l^m for the spherical harmonics of l-th degree and m-th order, the distribution can be written as $E_r(r)Y_l^m(\vartheta,\varphi)$. The radial component $E_r(r)$ can be written as [4]

$$E_r = \begin{cases} Aj_m(k^{(m,l)}n_1 r) & r \le R_1 \\ Bj_m(k^{(m,l)}n_2 r) + Ch_m^{(1)}(k_\varphi^{(m,l)}n_2 r) & R_1 < r \le R_2 \\ Dh_m^{(1)}(k^{(m,l)}n_3 r) & r \le R_2 \end{cases} . \tag{3.1}$$

Here, j_m and h_m are the spherical Bessel function and spherical Hankel function of the first kind, both of the m-th order. By setting equal refractive indices $n_1 = n_2$ for both sides, the equation directly describes a solid microsphere.

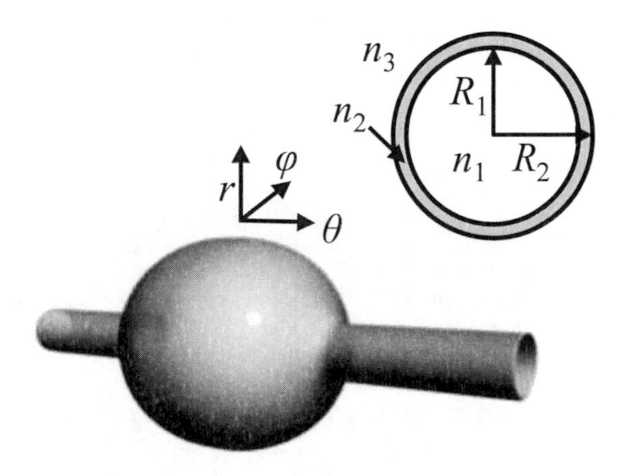

FIG. 3.1. Schematic of the general microbubble design (from [4]). For the analytic description, a spherical coordinate system with origin inside the microbubble is used.

A very interesting property of hollow microresonators is the possibility of filling the inside of the resonator with an arbitrary gas or liquid. For thin-walled microbubbles, the interaction of these inner media with the inner evanescent field of the WGMs can be quite strongly. In this case, the electric field distribution of higher order modes is mainly located inside the interior and the sensing capabilities of the devices are highly enhanced (see FIG. 3.2).

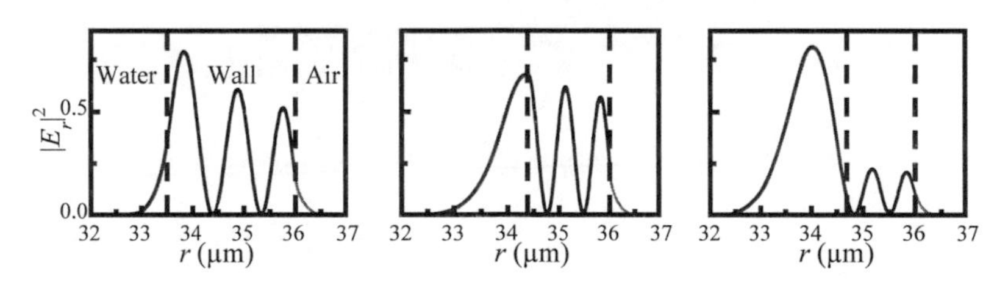

FIG. 3.2. Electric field distribution $|E_r|^2$ in the radial direction for the 3rd order WGM in microbubbles with different wall thicknesses (from [4]). It can be seen that for thin walls most of the field is located inside the inner volume.

3.3 Manufacturing

Although the manufacturing of simple solid spherical microresonators directly made from standard glass fibers or tapered silica rods is well-known for a long time [10], the production of hollow microspheres is a relatively new variant which was developed just over the last few years. For manufacturing solid microspheres, the most common technique is to thermally heat a very thin tip of glass or silica by a CO_2 laser beam. By carefully melting the material, a spherical droplet forms due to surface tension. This effect is also responsible for the high quality factors (Q factors) of the WGMs typically observed within such systems. Due to the tensioning process, the surface roughness of such spheres is highly reduced and reaches quadratic mean values below 0.5 nm [11]. In contrast to those solid systems, the fabrication of microbubbles is more critical. For a high reproducibility of the produced structures, exact and simultaneous control over the heating and pressurizing processes is required [12].

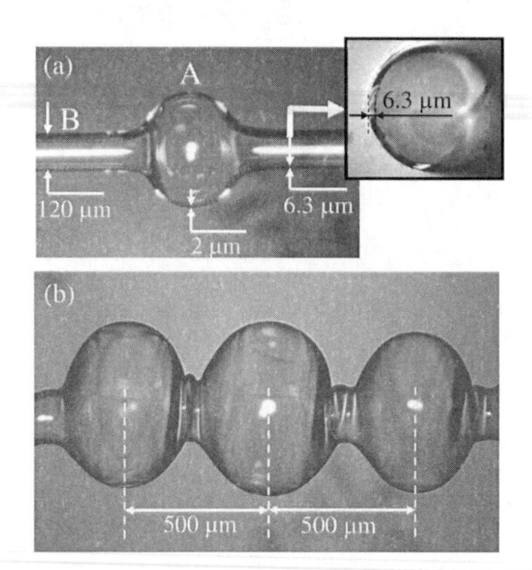

FIG. 3.3. Different types of microbubbles produced by Sumetsky *et al.* (from [13]). (a) Optical microscope image of a single microbubble. (b) Optical microscope image of three successive microbubbles.

The first realization of a microbubble was presented by Sumetsky *et al.* in 2010 [13]. For manufacturing, micropipettes were pulled in a fiber-drawing station from large silica tube preforms. The finally drawn micropipettes had diameters of d_{cap} = 120 µm and wall thicknesses of w_{cap} = 6.3 µm. The micropipettes were then mounted into a setup which allows simultaneous turning and pressurizing of the capillaries. A CO_2 laser beam is used to locally heat the micropipettes at their mid-section while they are constantly turned for homogeneous processing. Compared to classical convective heat sources, e.g., an oven or a gas torch, the advantage of a CO_2 laser beam is the very well defined heat spot. Also the higher power stability of the involved radiative heating process is an important aspect. The heating process is completely unaffected by air currents or heat streams as they would disturb an unprotected gas torch setup. While heating the mid-section of the micropipettes, the inside of the turning capillaries can be pressurized by a gas. As this mid-section is slightly heated above its glass transition temperature, the inner pressure can act on the softened part of the walls and locally expand the material. This process rapidly starts when the laser power is carefully increased to reach the glass transition temperature level of the material. The described manufacturing process is self-terminating. After reaching a specific wall thickness, the expansion suddenly stops and a stable microbubble is formed (see FIG. 3.3a).

This self-terminating process can be fully explained by the laser heating mechanism. The heating of the capillary is caused by just a small part of the energy within the laser beam. It must have been previously absorbed by the glass material. At constant laser power, this absorption is limited by the available amount of glass. The heating process depends on the effective volume of the absorbing material and so it is directly related to the wall thickness of the capillary. For thin capillary walls, the conveyed amount of energy is rapidly decreasing. In contrast, convective cooling of the structure only depends on the surface area and is thus independent on wall thickness. In the heated section, the ratio between material volume and surface area constantly decreases due to the expansion process. At a certain laser power, both processes become equal and a steady-state is established. This instantly freezes the further expansion and terminates the manufacturing process. The two concurrent processes of heating and cooling are illustrated in FIG. 3.4a.

Due to the high locality of the heating process, whole series of microbubbles can be produced along a single capillary. For instance, in FIG. 3.3b, a series with a central distance of 500 µm between consecutive resonators is shown [13]. The shape and size of the microresonators can be controlled by simply changing the laser power or the shape and size of the laser focus. Also the pressure inside the capillary can be changed to modify the expansion process. Another option to control the size of the produced microbubbles is to use pipette preforms with different aspect ratios between diameter and wall thickness.

The microbubbles produced by Sumetsky *et al.* had an average diameter of d = 370 µm and a calculated wall thickness of w = 2 µm [13]. In these cavities, WGMs with optical Q factors exceeding 10^6 could be observed. The presented method shows a good reproducibility and allows the controlled fabrication of various resonator sizes. The produced structures exhibit two openings or ports on the opposite sides of the microbubbles. The residual micropipettes can directly be used for microchanneling chemical or biological specimens. By passing also the interior of the microbubble, these specimens can interact with the evanescent fields of the WGMs inside the resonator and a continuous monitoring of the flow inside the microchannel becomes possible in detection-based applications.

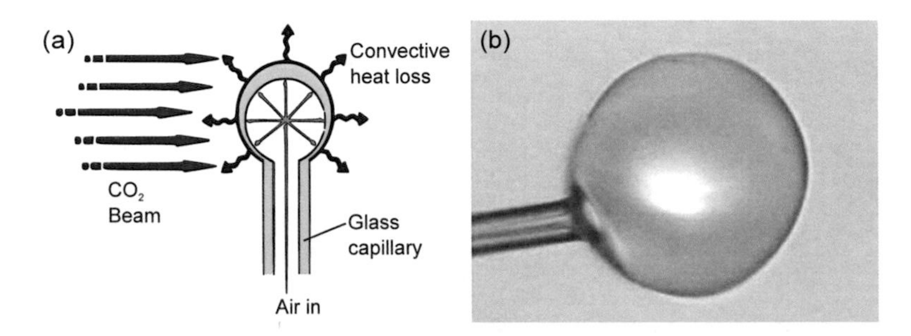

FIG. 3.4. (a) Schematic of the microbubble formation process due to simultaneous radiative heating and convective cooling. For moderate air pressures, a stable steady-state can be achieved. In this case, the expansion process is self-terminating. (b) A large microbubble viewed from the side by an optical microscope (from [14]).

In contrast to the described manufacturing method, a simplified technique without a rotational mount was introduced by Watkins *et al.* in 2011 [14]. The microbubbles produced with this method are only accessible from one side. However, this technique allows the production of smaller and thinner structures compared to the method presented by Sumetsky *et al.* [13].

A standard borosilicate capillary (d_{cap} = 1 mm, w_{cap} = 0.25 mm) was tapered down in a microburner driven drawing station (see FIG. 3.5a). The resulting microcapillary had an outer diameter of d_{cap}' = 25 µm and a wall thickness of w_{cap}' = 6.25 µm. One side of the capillary was closed by melting the tip within the focus of a CO_2 laser beam. The other end was directly connected to a pump. Thus, the inside of the capillary was pressurized with air. Afterwards, the closed end was again heated in the focus of CO_2 laser beam. By softening the glass, a spherical microbubble formed due to the internal pressure. Depending on the laser power and applied air pressure, an accelerated expansion could be observed. At moderate laser powers, this expansion suddenly stops as explained due to the reduced heat absorption inside the thin wall of a microbubble.

This method is not as reproducible as the method described by Sumetsky *et al.* [13], but it allows the manufacturing of smaller microbubbles. With this method, resonators with diameters $d < 100$ µm and wall thicknesses $w \approx 500$ nm could be achieved. The Q factors of the WGMs could be measured to be in excess of 10^5. However, the fabricated structures often show strong aberrations from a perfect spherical symmetry. In such cases, the resulting wall thickness becomes spatially dependent due to the likely oval shape of the resonators. It was shown that in such kinds of deformed microresonators Fano-type resonances can be observed [14].

The microbubbles in this thesis are based on a further development of the CO_2 laser setup used by Watkins *et al.* [14]. For the manufacturing process, the same type of borosilicate capillaries was used. They were also tapered down to micrometer size in a drawing station and then sealed on one end by the described CO_2 laser treatment. However, after pressurizing the capillary, the fabrication of the microbubbles does not take place on the top but at the mid-section of the microcapillary. This is in accordance to the method described by Sumetsky *et al.* [13]. But instead of turning the capillary, a homogenous heating profile was achieved by softening the glass in the common focus of two intersecting CO_2 laser beams. In FIG. 3.5b, a simplified schematic of the used microbubble manufacturing setup is presented. The resulting microbubbles are more symmetric compared to the published results of Watkins *et al.* [14]. In contrast to the manufacturing process of Sumetsky *et al.* [13], the presented technique allows to create microbubbles with much smaller diameters of below 100 μm and sub-micrometer wall thicknesses.

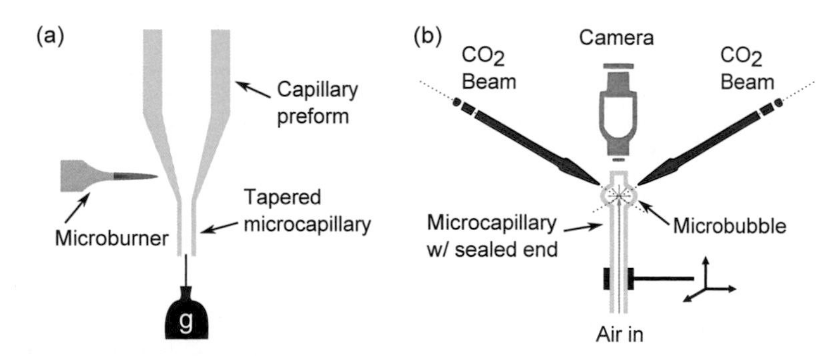

FIG. 3.5. (a) Microcapillary pulling process. A weighted borosilicate capillary is locally heated in the flame of a gas microburner and then tapered down to micrometer size by gravitational force. (b) Schematic of the microbubble manufacturing setup. The pressured microcapillary is heated from two sides to increase structural homogeneity. The intersecting CO_2 laser beams are tightly focused into the material.

In FIG. 3.6, examples of different optical microbubble resonators are shown. The structures produced with the two-beam manufacturing setup are normally single-ended. However, after manufacturing, it is also possible to reopen the closed part by shortly heating the pressurized sealed end with an intense laser pulse. The resulting two-port microbubbles are useful for experiments where open access to both sides of the cavity is required. The resulting microchannel allows constantly flushing gaseous or liquid media through the system and is useful for microchanneling applications were a high exchange rate of the inner cavity volume is required. An example of such a two port microbubble produced with the two-beam laser setup is shown in FIG. 3.6g.

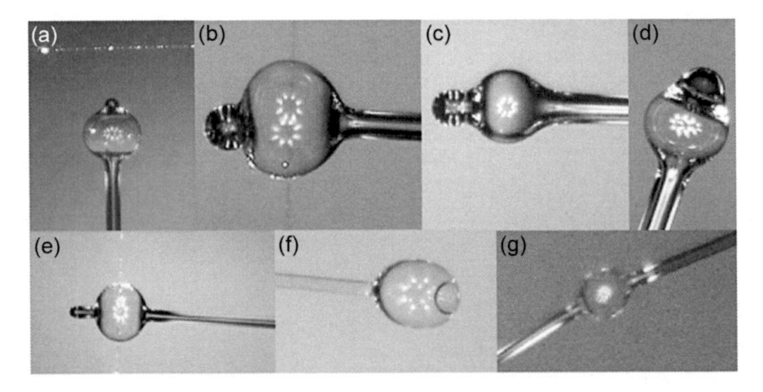

FIG. 3.6. (a)-(g) Examples of different microbubble resonators. In (a), (b) and (e) also a tapered fiber coupler can be seen next to the microbubbles. Light scattered from the tapered fiber is also visible. (f) A microbubble filled with water. The resonator was made by heating the capillary head on with a single CO_2 laser beam. (g) A microbubble with two openings or ports. The diameters of the presented resonators are (a) 265 μm, (b) 490 μm, (c) 170 μm, (d) 290 μm, (e) 330 μm, (f) 195 μm, (g) 95 μm.

3.4 Wall Thickness Estimation

For a precise estimation of the modal and pressure tuning properties of microbubble resonators, information about the diameters and wall thicknesses is required. The outer diameter of a microbubble can directly be measured by simple optical microscopy. This also holds for measuring the inner diameter in microbubbles with a sufficiently large wall thickness. The method does not work on all produced samples. For thinner walls, other techniques have to be applied. However, in some cases, direct approaches were still possible on accidentally broken microbubbles via scanning electron microscopy. For all other samples, a simple estimation method was developed and is presented in the following section.

The outer dimensions of the produced microbubbles could be directly measured by an optical USB digital measuring microscope (Reflecta Digiscope) which was integrated in the manufacturing setup (see FIG. 3.5b). For microbubbles with sufficiently large wall thickness ($w > 2$ μm), also the inner diameter could be determined by this method.

However, a direct measurement of the wall thickness in microbubbles is difficult with an optical microscope in the case of thin walls ($w << 2$ µm). On some samples, it was still possible to discriminate between the inner and outer wall by finding the right illumination conditions to increase the contrast. In these cases, a direct measurement of the two size parameters a and b, the outer and inner radii of the microbubble, was still possible.

However, if a direct measurement was not possible, an estimation method was used to determine the inner microbubble radius. It is assumed that during the manufacturing process of the microbubbles the CO_2 laser beam locally heats up just a small portion of the borosilicate. Within this region, the applied expansion pressure acts more or less spherical symmetric onto the inner surface. Then, due to a simple mass conservation approximation, the cross-sectional area of this region can be assumed to be invariant. For a standard capillary, this cross-sectional area can be described by a circular ring with outer capillary diameter D_1 and inner capillary diameter D_2. While tapering the capillaries for producing the microbubble preforms, the ratio between these values is constant. Therefore, the ratio between the two diameters D_1 and D_2 stays constant, too. Any additional material movement in the direction of the capillary axis is neglected for the estimation. The inner capillary diameter D_2 is completely determined by this ratio and the actual outer diameter D_1 such that no further measurement is needed after a tapering process. The cross-section of a microbubble manufactured from this tapered capillary preform is a circular ring which is completely defined by its outer and inner radii. Thus, the inner radius b of a microbubble can be calculated from the relation for the material area invariance by using the capillary diameters D_1 and D_2, and the outer microbubble radius a with

$$b = \sqrt{a^2 - \left(\frac{1}{4}D_1^2 - \frac{1}{4}D_2^2\right)}. \tag{3.2}$$

The wall thickness w of microbubbles can then be calculated as the difference between a and b by

$$w = a - \sqrt{a^2 - \frac{1}{4}D_1^2(1-\xi^2)}, \tag{3.3}$$

where $\zeta = D_2 / D_1$ is the constant ratio between the inner and outer diameters of the capillary preforms [8].

The estimated wall thicknesses calculated with Eq. (3.3) are in good agreement with the values measured by direct microscopy. For ultra-thin wall thicknesses, the validity of the size approximation could also be verified by scanning electron microscopy (SEM). As some of the produced microbubbles accidently broke into halves, it was possible to use SEM for a direct wall thickness measurement from the top (see FIG. 3.7). The measured results perfectly validate the approximation method.

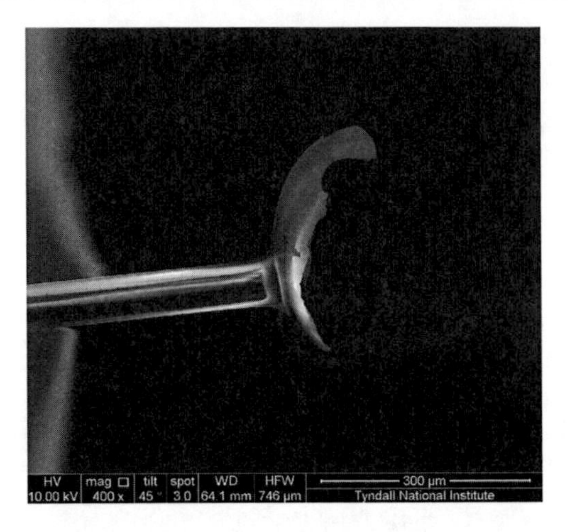

FIG. 3.7. Scanning electron micrograph of an accidentally broken microbubble. By this, the validity of the used microbubble size approximation was verified in the case of ultra-thin wall thicknesses ($w >> 1$ µm). For the presented microbubble resonator, a wall thickness of around 450 µm could be measured. This is in good agreement with the theoretically estimated value of around 500 µm.

Sumetsky *et al.* [13] presented an alternative method to estimate the wall thickness of microbubbles from directly measurable parameters. The microbubble wall thickness is also here estimated from considering simple geometric relations. By measuring again the diameter and wall thickness of the preform, the wall thickness of the resonator is calculated by

$$w_{Sumetsky} = (D_1 - D_2)\frac{D_1}{2a}. \tag{3.4}$$

For most of the produced microbubbles, the calculated wall thicknesses are of the same order as the results derived with Eq. (3.3).

A further approach for estimating the wall thickness of microbubbles was presented by Seifert *et al.* [15]. Here, the thickness of the wall is calculated by

$$w_{Seifert} = \frac{D_1}{8a}. \tag{3.5}$$

Also this equation is based on considerations about the area, but other assumptions for the size parameters before the expansion process are used.

All three given approaches are based on different assumptions about mass conservation during the expansion process. However, for many microbubbles, the actual differences between the three estimation methods are negligible. For microbubbles produced with the described two-beam manufacturing setup, Eq. (3.3) turned out to be the most accurate approach. It depends on the applied manufacturing process which of the three equations is the best for estimating the wall thicknesses of produced microbubbles.

3.5 Pressure Tuning

The microbubble resonators which were presented in the previous sections are all based on tapered microcapillaries made from borosilicate glass. These capillaries allow the production of microbubbles by simply pressurizing one end of the capillary. The other capillary end is sealed by melting the pipette tip within the focus of a CO_2 laser beam. During the microbubble manufacturing process, the pressure expanded parts of the capillary have to be heated by a laser beam or another suitable heat source, e.g., a burner flame or an electric heating element. The same expansion method can also be applied to the final resonators without heating the capillary. In this case, pressurizing the gas inside the microbubble causes a reversible deformation in the spherical shape of the cavity. This effect can be applied for tuning the WGM resonance of a microbubble with high accuracy.

In the following section, a general physical description of the tuning properties in spherical microbubble resonators is presented. It is based on basic assumptions about the symmetry of the system and takes the elastic properties of the resonator material into account. The derived formula is then applied to the specific borosilicate material of the used microresonators as they were presented in the previous sections. By that, a generalized simplified relation between the two principal diameters of a borosilicate microbubble and the final pressure tuning abilities is derived.

3.5.1 Analytic Model of WGM Shifts

The circulating waves in a WGM microresonator have to fulfill a resonance condition for constructive interference (see also Chapter 2.2.1). Hence, in the ray optics picture the resonance wavelength can be described by

$$2 \cdot \pi \cdot r_{eff} \cdot n_{eff} = m \cdot \lambda . \tag{3.6}$$

Here, r_{eff} and n_{eff} are the effective radius and the effective refractive index of the resonator, $m >> 1$ is an integer number corresponding to the number of field maxima around the circumference, and λ is the wavelength of the circulating light.

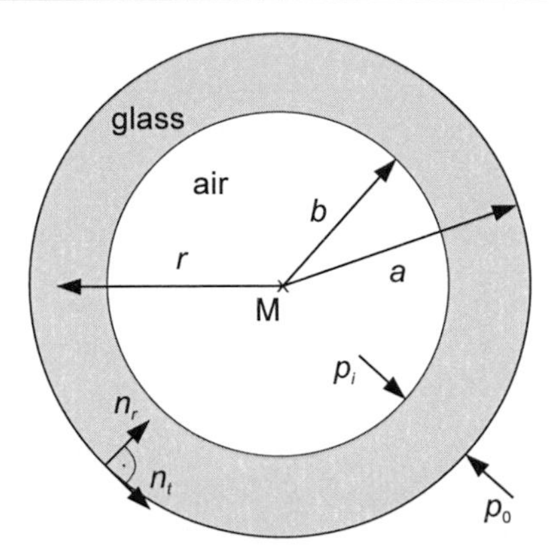

FIG. 3.8. Microbubble resonator with spherical coordinate system. a and b are the outer and inner radii of the shell, p_0 and p_i the applied external and internal uniform pressures. The refractive indices in the tangential and radial direction are denoted by n_t and n_r, respectively.

The relative shift of the resonance wavelength λ of a WGM is given by differentiating Eq. (3.6) and thus it can be written as

$$\frac{d\lambda}{\lambda} = \frac{dr_{eff}}{r_{eff}} + \frac{dn_{eff}}{n_{eff}}. \tag{3.7}$$

By changing the pressure inside the microbubble, the sphere will expand and induce changes to both the effective radius and effective refractive index of a mode due to stress and strain effects. A microbubble resonator is considered as a thin spherical glass shell with sufficiently large wall thickness (w > 2-3 λ) and size ($r_{eff} >> \lambda$) such that these two parameters are not directly linked to each other. For such a microbubble, also the effective values become identical to their absolute values and thus the index $_{eff}$ can be suppressed in the following discussion.

Let a and b denote the outer and inner radii of the spherical glass shell, and p_0 and p_i the applied external and internal uniform pressures (see FIG. 3.8). For the assumed spherical symmetry, the shear-stress and shear-strain components within the shell can be neglected. As the applied pressures are acting perpendicular and homogeneous to the surface, the symmetry of the problem is maintained. In this case, the problem can be expressed in spherical coordinates and the tangential component of the displacement vector vanishes for the compression and expansion of the microbubble. The solution of the well-known elasticity equation [16] then yields to a distribution of the principal stresses in radial r and tangential t direction, and to a term for the radial displacement u_r of infinitesimal elements inside the shell which are given by

$$\sigma_r(p_i) = \frac{a^3(r^3 - b^3)}{r^3(b^3 - a^3)} p_0 + \frac{b^3(a^3 - r^3)}{r^3(b^3 - a^3)} p_i \tag{3.8}$$

$$\sigma_t(p_i) = \frac{a^3(2r^3 + b^3)}{2r^3(b^3 - a^3)} p_0 - \frac{b^3(2r^3 + a^3)}{2r^3(b^3 - a^3)} p_i \tag{3.9}$$

$$u_r(p_i) = \frac{r}{6GK}\left[\frac{a^3(4Gr^3 + 3Kb^3)}{2r^3(b^3 - a^3)} p_0 - \frac{b^3(4Gr^3 + 3Ka^3)}{2r^3(b^3 - a^3)} p_i\right]. \tag{3.10}$$

Here, G and K are the shear and bulk modulus of the microbubble material. With these equations, the two summand terms which describe the relative shift of the resonance wavelength in Eq. (3.7) can be derived.

Pressure dependence of the radius

For evaluating the pressure expansion of the microbubble, Eq. (3.8) is evaluated at position $r = a$ and the strain term da/a derives as function of p_i and p_0 as

$$\frac{da}{a} = \frac{4p_0Ga^3 + 3p_0Kb^3 - 4p_ib^3G - 3p_ib^3K}{12GK(b^3 - a^3)}. \tag{3.11}$$

This equation describes how the outer radius a of the spherical shell is changed by varying the relation between the inner and outer pressure of the sphere. Thereby, the unstressed value for the radius a relates to a condition where both the inner and outer pressure are equal and vanishing.

Pressure dependence of the refractive index

For calculating the pressure dependence of the refractive index, the Maxwell-Neumann equation can be used [17]. For an isotropic and stress-free material, it follows that the refractive index changes in the radial r and tangential t direction are given by

$$n_r(p_i) = n_0 + C_1\sigma_r(p_i) + C_2(2\sigma_t(p_i)), \tag{3.12}$$

$$n_t(p_i) = n_0 + C_1\sigma_t(p_i) + C_2(\sigma_r(p_i) + \sigma_t(p_i)). \tag{3.13}$$

Here, C_1 and C_2 are the elasto-optic constants and n is the isotropic refractive index of the stress-free material. Due to the difference in the principal stresses between the tangential and radial direction, also the refractive index vary between these two directions. This induces a pressure induced birefringence within the material for light traveling in different directions. For evaluating the changes in the tangential direction of WGM light propagation, Eqs. (3.8) and (3.9) are inserted into Eq. (3.13) and again evaluated at position $r = a$,

$$\frac{dn_t}{n_t} = \frac{2C_1 p_0 a^6 + C_1 p_0 a^3 b^3 - 3C_1 p_i b^3 a^3 + 4C_2 p_0 a^6 - C_2 p_0 a^3 b^3 - 3C_2 p_i b^3 a}{2a^3 n_0 (b^3 - a^3)}. \tag{3.14}$$

In Eq. (3.14), the dependence of the tangential component of the refractive index is evaluated as relation between the inner and outer aerostatic pressures p_i and p_0. This is independent of geometrical changes and solely a description for the pressure induced change in the optical properties of the material. The stress-free refractive index again corresponds to a condition with equal zero pressure for the in- and outside of the shell.

3.5.2 WGM Shift in Borosilicate Microbubble Resonators

The pressure dependence of the WGM resonance shift in borosilicate microbubbles can be derived by adding the two upper equations. The found result can be simplified by considering the specific properties of the microbubble material. For the presented pressure tuning experiments, borosilicate glass capillaries from Schott (Schott Duran) were exclusively used. All the required parameters for that type of borosilicate glass are well-known and given by the manufacturer (G = 26.67·10^9 Pa, K = 3.55·10^{10} Pa, $\upsilon = 0.2$, $n_0 = 1.4712$, $C = 4\cdot10^{-12}$ m^2/N). For this specific material the two elasto-optic constants C_1 and C_2 are equal and thus $C_1 = C_2 \equiv C$ can be substituted in the equations. With $dn_t/n_t = dn/n$ and using $K = [2G(1+\upsilon)]/[3(1-2\upsilon)] = (4/3)G$, the summation of the two Eqs. (3.11) and (3.14) then gives a simple linear expression for the pressure induced resonance shift in borosilicate microbubbles,

$$\frac{d\lambda}{\lambda} = -\frac{d\nu}{\nu} = \frac{n_0(a^3 + b^3) + 12CGa^3}{4Gn_0(b^3 - a^3)} p_0 - \frac{2n_0 b^3 + 12CGb^3}{4Gn_0(b^3 - a^3)} p_i. \tag{3.15}$$

In FIG. 3.9, the pressure tuning curve of a typical borosilicate microbubble is presented together with the tuning curve based on the pure geometrical change due to pressure expansion. It can be seen that with increasing inner aerostatic pressure the resonance frequency linearly shifts to lower frequencies. Thereby, for both curves, the pressure expansion curve as well as the pressure dependent refractive index curve, the WGM resonance frequencies are shifting into the same direction. This can be physically explained by simply interpreting a WGM as an interfering wave around the spherical shell. By expanding the radius of the cavity, the wavelength of a specific resonance shifts to longer wavelengths, i.e., to lower frequencies. For the refractive index induced shift, a similar interpretation can be used. As the stress term increases the effective value of the refractive index, the wavelength of the circulating light has to be increased for sustaining a specific resonance. For the presented microbubble, the influence of the two shifting components is nearly equivalent.

The pressure induced shifting properties are independent on the outside pressure. It just defines the intercept of the shifting curve compared to a stress-free microbubble with equivalent uniformly applied inner and outer pressures. By ignoring the constant term, Eq. (3.15) simplifies to

$$\frac{d\lambda}{\lambda} = -\frac{d\nu}{\nu} = \frac{2n_0 b^3 + 12CGb^3}{4Gn_0(b^3 - a^3)} p_i . \tag{3.16}$$

As for this equation only geometric and material changes are considered, a possible influence of pressure tuning on the WGM intensity distribution is neglected. In the case of very thin wall thicknesses ($w \approx \lambda$), a substantial portion of the circulating wave is guided outside the silica. These evanescent parts of the WGM are strongly leaking into the inner and outer environment of the shell. When pressurizing the microbubble, the mode structure can also be affected by a change of the inner volume. The resulting field distribution may be substantially altered if compared to a thicker walled microbubble where most of the field is capped inside the silica and thus protected against changes along the inner boundary. For considering these effects, the refractive index n_0 has to be replaced by the stress-free effective refractive index n_{eff} of the corresponding WGM. This value can be obtained by using a suitable numerical simulation software, e.g., an optical mode solver (see also Chapter 2.4 and Appendix A), or by an analytic approach based on the results given in Chapter 2.3.2.

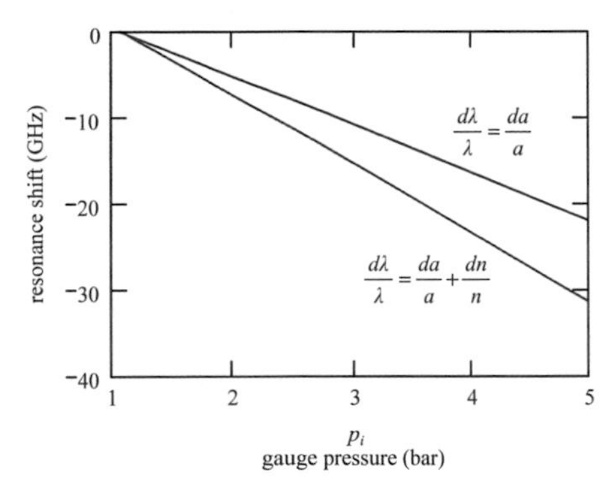

FIG. 3.9. Pressure induced resonance shift of a typical microbubble made from borosilicate glass (D = 100 µm, w = 2 µm, λ = 770 nm). The upper curve corresponds to the pure pressure expansion while the lower curve also includes the change in the refractive index. The outer ambient pressure is 1 bar.

In addition to an altered profile by 'squeezing' the mode out of the shell, the evanescent parts of the circulating wave may also show a stronger interaction with media at the inner and outer sides of the microbubble. Thus, for ultra-thin wall thicknesses ($w << \lambda$), also the effect of the pressurized inner volume must be considered for the resulting resonance shift. In this case, the effective refractive index n_{eff} of a WGM becomes dependent on changes in the refractive index of the inner volume. In the presented measurements, the influence of the wall thickness could be neglected and thus a detailed analysis of the mentioned effects is not presented.

The relation between the resonance shift and the applied internal pressure as derived in Eq. (3.16) allows a further simplification by inserting the corresponding values for C, G, and n_0. It follows that the resonance shift in a microbubble is directly proportional to a geometric parameter $\chi = a^3/(a^3 - b^3)$ and therefore to the ratio between the whole enclosed volume and the volume of the spherical shell alone. With the given material parameters of the used borosilicate capillaries, the theoretical proportionality factor between the slope of the resonance shift and the geometric parameter χ can be calculated as approximately −1.1 GHz/bar. Thus, the WGM shift in borosilicate microbubble resonators due to pressure tuning can be written as

$$\frac{d\lambda}{\lambda} = -\frac{d\nu}{\nu} = -1.1\frac{\text{GHz}}{\text{bar}} \cdot \chi \cdot p_i \cdot \tag{3.17}$$

With this relation, it is possible to estimate the pressure tuning abilities of produced microbubbles directly from geometric considerations by simply measuring the principal radii a and b.

3.6 Pressure Tuning at Room Temperature

In the following, the presented analytic model for the pressure tuning abilities of spherical glass shells is applied to pressure tuned borosilicate microbubbles at room temperature. Thereby, the validity of the used theoretical assumptions is verified by experimental data and the possibility of pressure tuning microbubbles over a full resonator FSR is demonstrated.

3.6.1 Experimental Methods

For the experimental verification of the theoretical tuning model presented in the last section, borosilicate microbubbles with outer diameters between 100 µm and 500 µm were produced by the methods described in Section 3.3. For these microbubbles, different wall thicknesses ranging between 1 µm and 7 µm could be realized. The measurements were performed either by direct optical microscopy or by applying the estimation method presented in Section 3.4.

The microbubbles were made from single-ended borosilicate capillaries. They were produced near the end of the closed side, not directly on the melted tip. The length of the residual section was chosen without any obligation. It was only ensured that the weight of this section did not deform the later microbubble during the manufacturing process. An example of a typical microbubble is shown in FIG. 3.10a.

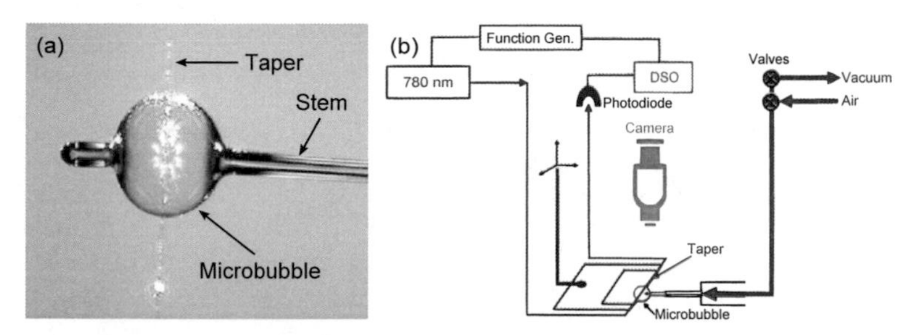

FIG. 3.10. (a) Example of a typical microbubble used for the pressure tuning experiments. The outer diameter of this bubble was 170 μm having a wall thickness of 1.5 μm. (b) Schematic of the setup with laser control, camera, digital storage oscilloscope (DSO), and air supply system.

The WGMs in such a microbubble were observed by a standard coupling setup as schematically shown in FIG. 3.10b. A tunable external cavity diode laser (New Focus Velocity) emitting at wavelengths around 780 nm (see also Appendix B) was used for optically probing the resonances. Thereby, the laser power was kept as low as possible (below a few 100 mW) to avoid any optical heating or cooling of the resonator while in near resonance with the laser emission. The laser wavelength could be frequency detuned within a mode-hop-free scanning range of 60 GHz. By applying a 5 Hz saw tooth ramp signal from an external frequency generator, continuous scanning over a single resonance or a small set of resonances was possible.

The laser light was coupled into a tapered optical fiber as a simple coupling device to excite different WGMs inside the microbubble. The waist diameter of this fiber was around 1 μm allowing only a few spatial modes to propagate. The coupler was in full contact with the resonator and thus the system was in the strongly over-coupled regime throughout the measurements. Although this is reducing the measured Q factors of the system, the full contact operation ensures stable coupling conditions during pressure tuning. This is highly important as otherwise the induced geometrical changes would require a continuous recoupling throughout the measurement series.

After passing the resonator coupling zone, the transmitted laser light, which exits the fiber taper, was out-coupled and detected by a fast silicon photodetector (Thorlabs DET36A). The different cavity resonances could be observed as Lorentzian shaped dips in the measured optical power (see inset of FIG. 3.11). For their detection, a digital storage oscilloscope (Tektronix TDS2022C) and an analog-to-digital data acquisition card (NI PCIe-6321) were used. In this experimental setup, the examined microbubbles showed Q factors ranging between 10^3 to 10^7 depending on the wall thickness and the symmetry of the microbubbles. At thicknesses below 1 μm, the number of observable modes is highly reduced because of the low material volume. For such thin structures, also the measured Q factors are degraded. This is due to an increased interaction of the modes with the inner and outer surfaces of the cavity. However, by scanning the laser frequency over the full 60 GHz scan range, typically still a multitude of different modes can be observed.

For pressure tuning the microbubbles, the exit port of the capillary was connected to a compressed air line with integrated pressure control. The maximum gauge pressure was up to 6 bar. This was also the limit of the used gas supply system. Higher line pressures can be applied for a further extension of the available tuning ranges in the connected microbubbles.

When the air pressure inside the microbubble was increased, a red shift of the WGMs could be observed. This corresponds to the theoretical model presented in the last section. It predicts an increase in the physical dimensions of the resonator as well as an increase in the refractive index of the resonator material. During the measurements, the setup continuously monitored the spectral positions of the WGMs depending on the actual value of the applied pressure. For most of the analyzed microbubbles, the observed tuning range exceeded the available scan range of the laser. In these cases, when a WGM shifted out of the laser scan range, a corresponding mode at the start of this range was instead monitored. As these two modes are not necessarily belonging to the same mode family within one FSR, slightly different tuning properties may be observed. For example, higher order modes normally show a larger evanescent part compared to the fundamental mode. This larger part increases the interaction with the external environment. Hence, for higher order modes, this influence should not be neglected and the observed resonance shift of higher order modes may fractionally differ from the mode shift of the fundamental mode. However, nearly all of the observed modes showed a highly equidistant shifting behavior. In the performed measurements, the identical tuning behavior of two examined modes, which shifted in and out of the corresponding scan range of the laser, was carefully ensured by directly comparing their respective tuning rates.

3.6.2 Results

For measuring the pressure tuning properties, six microbubbles with different outer diameters and wall thickness were produced. In FIG. 3.11, the result of a corresponding measurement series is presented. The graph shows the observed resonance shift of a typical WGM against the applied inner gauge pressure. This shifting also describes the tuning properties for the majority of the observable modes and thus for the WGMs in general. However, for a few specific modes in these microbubbles, slightly different tuning properties were noticed. The reason for this difference is the modal dependence of the tuning properties as it was already explained in the last part of the experimental methods section.

In FIG. 3.11, straight lines are visible. These lines are the theoretical tuning curves for the examined microbubbles. They are results of the pressure tuning model introduced in Section 3.5. There is a good agreement between the specific fit lines and the respective measurement series. The theoretical model assumes a symmetric sphere with uniform wall thickness. It does not take into account dimensional irregularities as they may have arisen during manufacturing. It also does not consider any influence of the residual capillary or capillary stem regions. The agreement between the experimental and the theoretically derived tuning curves successfully proves the validity of the assumptions made in the model.

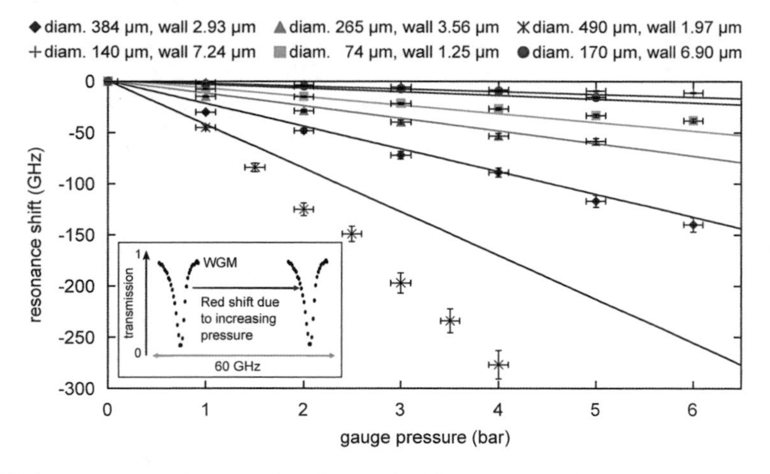

FIG. 3.11. Pressure tuning properties of WGMs in different microbubbles and comparison with the theoretical model as introduced in Section 3.5. The inset illustrates the measured Lorentzian resonance dip in the transmitted intensity and the pressure induced resonance shift.

However, it shall be noted that the theoretical model can only be seen as estimation for the resulting tuning properties of microbubbles. Due to geometrical variances from perfect spherical symmetry, the theory is not always giving the correct tuning ranges. For the larges microbubble examined within the presented measurement series, the observed tuning is much larger as predicted by the model. It fits for lower gauge pressures below 1-2 bar, but seems to double above these values. It is assumed that this difference is caused by asymmetries which can easily appear during the microbubble manufacturing process. For large microbubbles, the heat distribution, which is imposed by the CO_2 laser, may be inhomogeneous and thus the resulting wall thickness may be inhomogeneous, too. This was measured and verified by direct optical microscopy and may be the reason for the observed larger tuning range of this specific microbubble. However, also in these cases the theoretical model provides a good estimation for the lower bound of the available tuning range.

For the six examined microbubbles, very large tuning ranges could be observed. In case of large thin-walled microbubbles, maximum tuning ranges of around 300 GHz were measured for an internal gauge pressure of 6 bar. This tuning range may be further enhanced by starting with vacuum inside the microbubble or by applying higher gauge pressures. All the analyzed microbubbles, even those with thin wall thicknesses in the micrometer range, were able to withstand the maximum line pressure. It is assumed that much higher pressures can be applied, but previous testing of the mechanical stability is highly advised.

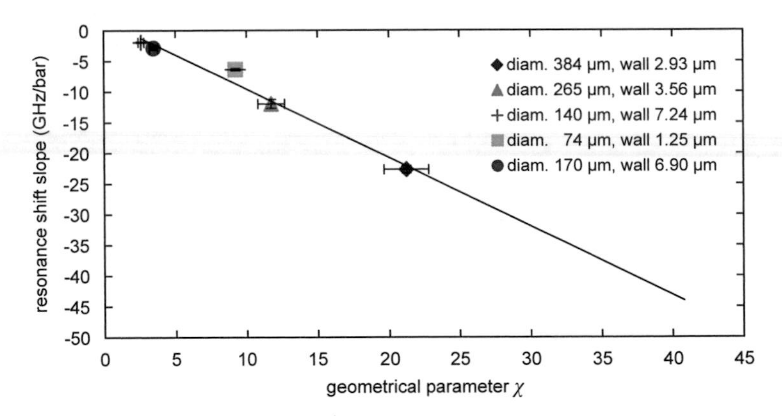

FIG. 3.12. Dependence of the average experimental pressure tuning slopes of the different microresonators presented in FIG. 3.10 with the geometric parameter $\chi = b^3/(a^3-b^3)$ as introduced in the theory section (Section 3.5.2). a and b correspond to the inner and outer diameters of the microbubble.

For the different measurement series shown in FIG. 3.11, the average slope obtained from the experimental data was calculated to further investigate the pressure tuning dependence of microbubbles. In FIG. 3.12, the result of this analysis is presented. The average slope of the largest microbubble, which not corresponds with the theoretical tuning model, was excluded from the analysis. In the figure, the average tuning slope is plotted against the geometric parameter χ (see Section 3.5.2). This parameter only depends on the principal geometry of the microbubbles and allows a simplified prediction of their pressure tuning abilities.

For the analyzed microbubbles, excellent agreement between the experimental data and the theoretical predictions is observed. For microbubbles produced from borosilicate capillaries, a proportionality factor between the slope of the resonance shift and χ is calculated as –1.1 GHz/bar. From the experimental data shown in FIG. 3.12, this factor can be measured as –(1.1 ± 0.1) GHz/bar. With this factor, the tuning range of borosilicate microbubbles, i.e., the resonance shift per applied internal uniform gauge pressure, can be directly predicted from Eq. (3.17). For this equation, only the inner and outer diameters of the produced microbubbles have to be measured without any further assumption.

The presented theoretical model was developed with borosilicate microbubbles in mind. By changing the capillary material and replacing the corresponding mechanical properties in Eq. (3.16), the observed linear behavior can also be transferred to other material systems. Thus, the general theory for the pressure tuning of microbubbles is verified by the experimental data in borosilicate cavities and a simplified method for estimating their tuning properties is given by introducing a geometric parameter χ.

3.7 Pressure Tuning under Cryogenic Conditions

The presented pressure tuning method offers a very simple but versatile technique to reversibly shift the WGM resonances in microbubbles. Besides the inner microbubble pressure, no other physical parameters must be considered. For sensing applications or by using WGM resonators as molecular detectors, this is a major advantage compared to other resonator tuning methods, e.g., by simple temperature tuning. Furthermore, as this tuning method is based on purely intrinsic microbubble properties, the resonance shift can be decoupled from the physical conditions of the external environment.

This makes pressure tuning a very appealing method for quantum optics applications under cryogenic conditions. Often the emission lines of applied photon emitters have to be perfectly matched to specific WGM resonances of a microresonator. Although these coupled systems can sometimes be analyzed at room temperature, in quantum optics cryogenic conditions are preferred due to the highly suppressed interaction with the environment, e.g., by reducing the decoherence rate of the applied photon emitters. In a cryostat, also the thermal bistability of a resonance can be suppressed and thus the intrinsic Kerr non-linearity of the material may become dominant [18]. Attempts to couple the narrow zero-phonon line of color centers in diamond to WGMs have been reported by several groups [19]–[21], but a required reliable tuning method could not be realized so far. Under cryogenic conditions, changing the circumference of a WGM resonator by temperature tuning via thermal expansion is not appropriate over large tuning ranges [22], [23]. Other tuning methods are based on mechanical strain, for example, by compressing or stretching microspheres via piezoelectric actuators [6].

By using internal aerostatic pressure for tuning the WGMs in a microbubble resonator, practically no additional thermal load is applied to the cooling system. This is an important prerequisite for cryogenic applications in quantum optics as highly stable temperature control of the experimental setups is required. By applying the internal microbubble pressure via a hermetically sealed gas support line, the influence on the physical environment of the resonator or the coupled system should be negligible. Nevertheless, when applying an optical microresonator to a cryogenic environment, the influence of precursory resonance shifts has to be taken into account. These shifts are caused by the negativity of the thermal expansion coefficient in the resonator material when cooling the system from room temperature to the required low temperature working range. By practically shrinking the physical dimensions of the resonator, all WGM resonance lines are shifted to smaller wavelengths. For a controlled matching between a specific WGM resonance line and the emission line of an applied photon emitter, this effect has to be considered during the cool down. When approaching the working temperature of the cryostat, at least a rough pre-matching of the two involved resonance lines should already have been established.

In the following section, the possibility to apply pressure tuning to microresonators in a cryogenic environment is investigated and their total tuning behavior is analyzed for cooling such a system from room temperature to liquid nitrogen (LN) temperature. The general thermal shift and tuning behavior of borosilicate microbubbles is theoretically modeled and compared to experimental data. It is shown that the stress/strain-based pressure tuning with compressed air or gas is widely unaffected by the actual system temperature.

3.7.1 Experimental Methods

The microbubbles used for the cryogenic pressure tuning experiments were widely identical to the microbubbles described in the experimental methods section for the pressure tuning properties under room temperature conditions (see Section 3.6.1). Most of the experiments were done with a microbubble having an outer radius a = 108 μm and an inner radius b = 105.6 μm. These physical dimensions correspond to a wall thickness of 2.4 μm and represent a typical microbubble which can be produced with the manufacturing methods described in Section 3.3. Concerning to the given theory of temperature tuning under room temperature conditions, this values refer to a geometric parameter $\chi = 14.3$ and result in a pressure tuning slope of –15.7 GHz/bar.

The principal measurement setup for exciting and analyzing the spectral resonance positions within the microbubble resonator was identical to the setup already described in the room temperature measurements section (Section 3.6.1). The only difference consisted in an additional home-build cryostat system. It enclosed the complete setup including the resonator and the applied fiber taper. In FIG. 3.13, a picture of the used microbubble and a schematic of the experimental configuration are shown.

The cryostat system was designed as open bath cryostat. A plastic containment was used as vessel for holding the cryogenic fluid, i.e., liquid nitrogen (LN) from a mobile Dewar system. The coupling and tuning setup was separated from the LN reservoir by a small metal box just large enough to support all required experimental components. Inside the metal box, a gas feedthrough was connected to a plastic tube which held the microbubble in place. After the bubble was inserted, the plastic tube was sealed by standard UV curing glue. For holding the tapered optical fiber coupler, a U-shaped simple plastic mount made from glass fiber reinforced thermosetting plastics (GRP) was installed. It was able to hold the fiber without losing the mechanical tension of the taper during system cool down. The fiber mount was connected to a three-dimensional translation stage by a post mount which was fed through the sealing lid of the metal box. The lid of that box was made from clear plastics thus that the position of the taper could be monitored by using an optical microscope and a camera. It was also possible to flush the inner metal box with pure nitrogen gas (N_2) before starting the cooling experiments. As this flushing highly reduces the amount of oxygen and other naturally appearing gases inside the containment, no icing effects on any of the active optical components could be observed. This is highly important due to the fact that under normal gaseous conditions the cold fiber taper and the resonator surface will cause an enhanced deposition and solidification of gases having a higher boiling point compared to nitrogen. Without flushing, any transmission through the fiber taper broke down due to the surface creation of ice when reaching temperatures below a few tens of Kelvin under 0 °C.

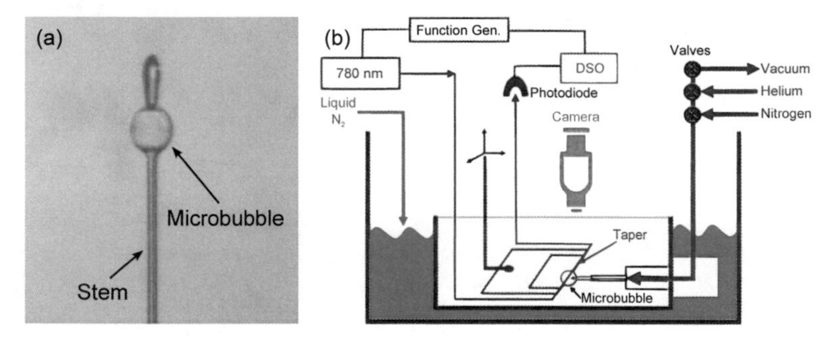

FIG. 3.13. (a) Microbubble used for the cryogenic experiments as seen through the microscope above the lid of the metal box. (b) Schematic of the cryogenic setup with laser control, camera, digital storage oscilloscope (DSO), and gas supply system.

For cooling the microbubble, the gap between the inner metal box and the plastic containment was filled with LN from the Dewar system. During filling and subsequent cooling, the inside of the metal box was flooded with N_2 to prevent icing on taper and microbubble. The actual temperature of the gas inside the metal box was continuously measured by an integrated thermocouple which was positioned next to the resonator. After sufficient cooling, the flow of N_2 could be stopped and the optical transmission of the fiber remained stable for the rest of the cool down process. The box temperature settled at 86 K and stayed at this value for some minutes as long as the LN boiled off. However, the settled temperature could be stabilized for even longer working times by simply adding more LN to the bath. When the LN boiled off to a certain level, the box started to slowly heat up. It reached room temperature again after fully evaporating the entire LN reservoir but without creating ice on any of the optical components.

In this way, it was possible to measure the temperature-driven WGM shift continuously from the cold to the hot state without requiring the stabilization of specific temperatures in between. The method of starting at low temperatures also offers more stability to the whole measurement process. There are no mechanical vibrations involved as they were associated with the off-boiling cryofluid or a necessarily required refilling of the LN reservoir. However, due to the different linear expansion coefficients of the various materials inside the box, coupling between the resonator and the fiber taper can still get lost sometimes during the heating process. In such cases, an instant manual recoupling was performed.

As already shown for the room temperature pressure tuning experiments described in Section 3.6.1, the large resonance shifts of the WGMs could be fully recorded by matching the shift rates of modes leaving the fixed frequency scan range of the laser system on the red side and new modes continuously appearing at the blue side of this range. As discussed for the tuning experiments performed under room temperature conditions, in the cryostat again some of the modes showed a small difference in their observed shift rates. This is due to a stronger interaction of these specific WGMs with the environment of the microbubble. However, the majority of modes are shifting at the same rate.

To analyze the resulting shift rates, a complete heating process was recorded by a data acquisition card (NI PCIe-6321). It also correlated the actual system temperature to the spectral resonance positions of the measured WGMs. Thereby, the timing resolution of our setup allowed a discrimination of half a second which was more than sufficient even at lowest system temperatures where the largest shift rates were observed. In the recorded data, specific modes were traced within the laser scan range to get an average value for the differential shift rate (DSR) at specific system temperatures. The total shifting behavior of WGMs from LN temperature to room temperature and vice-versa was then calculated by integrating these DSR values.

For the pressure tuning experiments, the microbubble was mounted within the setup and standard room temperature tuning was performed as described in Section 3.6.1. By this, the validity of the room temperature model could be verified also for the specific microbubble under investigation. Afterwards, the complete setup was cooled down and stabilized at the lowest achievable system temperature before a second pressure tuning experiment was performed.

In contrast to the room temperature experiments, in the used cryogenic setup, gaseous helium (He_2) instead of gas from a compressed air line was applied for pressure tuning the microbubbles. This change was required because nitrogen (N_2) in the air tends to be liquefied inside the microbubble when pressurizing at LN temperatures. In this case, a sudden phase transition could be observed which prevents stable measuring conditions. This is possibly due to an enhanced mechanical vibration or a sudden change in the effective refractive indices of the observed modes. These effects could be fully avoided by using helium for the pressure tuning process in the low temperature range. Due to its lower boiling point, such a phase transition can not be observed within the achievable temperature range of the cryogenic setup. However, the physical process of pressure tuning is not affected by a change of gas. An application for the liquefaction of gaseous nitrogen inside a cooled and pressurized microbubble could be experiments which aim to observe specific non-linear interactions between LN and WGMs. The change of the internal pressure tuning gas does not affect the observed tuning range of a microbubble. This independence on the used tuning gas could be verified by comparing the room temperature tuning ranges when using compressed air and He_2 for pressurizing. It was also ensured that no gas leakages occurred and that the thin wall of the microbubble was sufficiently helium-tight.

Before all measurements started, the inside of the microbubble was evacuated to avoid icing effects caused by residual gases which could possibly condensate during cooling. This also allows starting the pressure tuning process from 0 bar internal pressure. With the available helium gas supply system, a maximum inner gauge pressure of 4 bar was achievable. Under cryogenic conditions, typically another set of modes was observed compared to the room temperature measurements. This is mainly due to the different size of a resonator at different temperatures. However, the shift of the resonances does not dependent on a specific mode set and can be autonomously measured.

3.7.2 Temperature Tuning of Borosilicate Microbubbles

For the theoretical description of the temperature tuning properties of microbubbles, Eq. (3.7) was used. By considering the thermal dependencies of the two relevant physicals parameters, the WGM resonance positions at arbitrary system temperatures were estimated. The change in the physical microbubble dimensions $a(T)$ by varying the system temperature between room temperature and LN temperature is described by the linear thermal expansion coefficient $\alpha = \frac{1}{a}\frac{da}{dT}$. For the description of temperature dependence of the refractive index $n(T)$, the thermo-optic coefficient $\beta = dn/dt$ is relevant. With these two terms, Eq. (3.7) can be rewritten to find an expression for the temperature induced resonance shift in WGM resonators by [22]

$$\frac{d\lambda}{dT} = \left(\alpha + \frac{1}{n}\beta\right)\lambda . \tag{3.18}$$

For this thermal resonance shift, only material parameters have to be taken into account and no dependence on the absolute size of the resonator is involved. From elasticity theory, it can further be shown that the temperature induced changes in the physical dimensions are unaffected by the concrete resonator design [16]. Full microspheres and microbubble resonators just consisting of thin spherical shells show absolutely identical temperature tuning behaviors. As long as the wall thickness of a resonator is sufficient to fully support the mode volume of a WGM, the theory is completely independent on this property.

The linear thermal expansion coefficient α and the thermo-optic coefficient β can be found in the literature for arbitrary materials. However, these values are normally given as constants only valid at specific temperature ranges. To fully analyze the temperature dependence of the temperature tuning properties, it should be taken into account that these coefficients itself also depend on respective temperature ranges. For the refractive index n, this temperature dependence is normally tolerably small. In the following, this change is neglected and the refractive index is fixed to the room temperature value n_0. By introducing the residual dependencies in Eq. (3.18), the temperature dependence of the resonance shift can be written as

$$\frac{\Delta\lambda}{\lambda} = \int\left(\alpha(T) + \frac{\beta(T)}{n_0}\right) dT \,. \tag{3.19}$$

For the used borosilicate capillaries (Schott Duran), the thermal expansion coefficient at room temperature and the refractive index for a wavelength of 770 nm are given by the manufacturer ($\alpha = 3.3\cdot10^{-6}$ K^{-1}, $n_0 = 1.4712$). The temperature dependence of the thermal expansion coefficient $\alpha(T)$ can be found in the literature [24]. In FIG. 3.15a, the experimental measuring curve from this reference is shown. For the temperature range under investigation, a quadratic fit shows nearly perfect agreement with the given data. Thus, the referenced temperature dependence of the thermal expansion coefficient can be approximated by

$$\alpha(T) = (-3\cdot10^{-5}\cdot T^2 + 0.0213\cdot T - 0.265)\cdot10^{-6}\mathrm{K}^{-1} \,. \tag{3.20}$$

The room temperature value and the temperature dependence of the thermo-optic coefficient $\beta(T)$ could not be found in the literature. For borosilicate as a special type of industrial glass, no specific optical data is provided by the manufacturer. However, the temperature dependence can be deduced from looking at some other types of optical borosilicate glasses which are also offered by Schott. For such kinds of glasses, the parameters α and β often have the same order of magnitude and therefore for a rough estimate $\beta(T) = \alpha(T)$ can be assumed.

Based on this simple approximation, Eq. (3.19) allows calculating the differential shift rate $d\lambda/dT$ (DSR) for arbitrary temperatures. At room temperatures around 300 K, the

DSR is in the range of –2.2 GHz/K. For the lowest temperature which is achievable with the used cryogenic setup, a DSR of –1.0 GHz/K is found. For the fully analyzed microbubble, a total resonance shift of nearly 350 GHz is expected for cooling from room temperature to LN temperature. The resulting temperature tuning properties of borosilicate microbubbles are in good agreement with the tuning properties of ordinary silica microspheres [22]. As for them, the observable tuning slope is strongly decreased at lower system temperatures, but it strongly flattens at to room temperature conditions.

3.7.3 Temperature Dependence of Pressure Tuning

The standard room temperature pressure tuning properties of borosilicate microbubble resonators are approximately given by Eq. (3.16). For a theoretical description of the influence of a temperature change on these given tuning properties, the temperature dependence of the different material parameters has to be fully taken into account. For the refractive index, it can be shown that the pressure tuning properties are widely unaffected by the small refractive index change which is induced by cooling the resonator from room temperature to LN temperature. Therefore, the constant refractive index of the stress-free material at room temperature n_0 can be used for calculating the temperature dependence of pressure tuning.

Following Eq. (3.16), the expected resonance shift in microbubble resonators is directly proportional to a given geometric parameter $\chi = a^3/(a^3 - b^3)$. As this parameter only depends on the ratio between the whole enclosed volume and the material volume of the spherical shell alone, it is independent on a relative size change due to cooling and heating. This geometric property stays constant during any temperature changes. Thus, a temperature dependence of the pressure tuning abilities is only possible from the temperature dependencies of the material parameters $G(T)$ and $C(T)$ as they were also introduced with Eq. (3.16). These two variables are the mechanical shear modulus and the elasto-optic constant, respectively. At room temperature conditions, these values are well-known and given by the manufacturer (see Section 3.5.2).

However, the temperature dependence of these material parameters could not be found in the literature. For the shear modulus, it was instead possible to fully estimate this dependence from a comparable borosilicate glass made by another manufacturer (Corning Pyrex). In this type of glass, the elastic shear modulus was found to behave temperature-positive [25]. Following this reference, the elastic modulus $G(T)$ should slightly increase with lower temperatures, but differ not more than 2-3 % from its room temperature value. For the temperature dependence of the elasto-optic constant $C(T)$, no literature data were found. The known room temperature value was instead assumed as constant over the full accessible temperature range of the cryostat. Under these approximations, it can be estimated that the total cryogenic pressure tuning properties of microbubbles are more or less of the same order as they were observed under room temperature conditions.

3.7.4 Results

In FIG. 3.14, the experimentally measured temperature shift rates of WGM resonances in a borosilicate microbubble are presented. From these data, at room temperature a DSR of –2.5 GHz/K was observed. At LN temperature, this value is strongly decreased to roughly –0.6 GHz/K. By integrating the intermediate DSRs, a total shift of nearly –325 GHz could be observed for tuning from room temperature to LN temperature. All these values are in good agreement with the theoretical estimations which are based on the assumption that the temperature dependence of the thermo-optic coefficient in borosilicate glass can be neglected. Thus, the rough estimate of $\beta(T) = \alpha(T)$ is validated by this measurement.

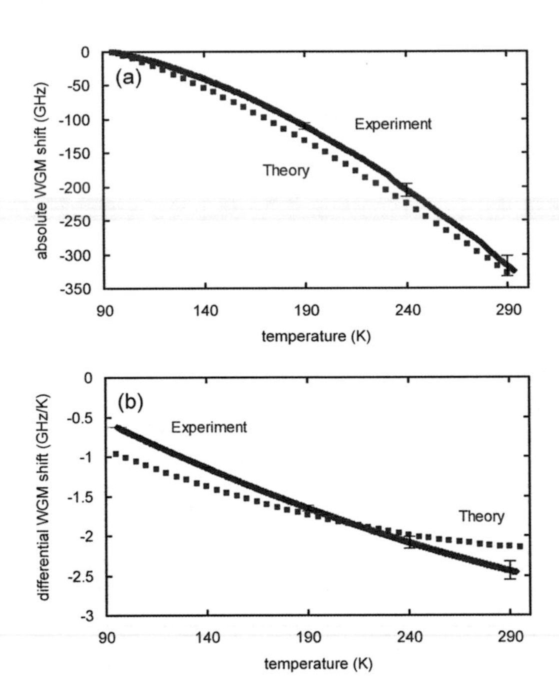

FIG. 3.14. (a) Absolute WGM resonance shift in a borosilicate microbubble by heating from LN to room temperature. The red curve represents the theoretical estimation based on the given assumption about the temperature dependence of thermo-optic coefficient while the blue curve is based on experimental data. (b) Differential shift rate (DSR) of the two curves. The red curve belongs to theory, the blue curves represents the experimental data. The errors are exemplarily indicated at elevated temperatures.

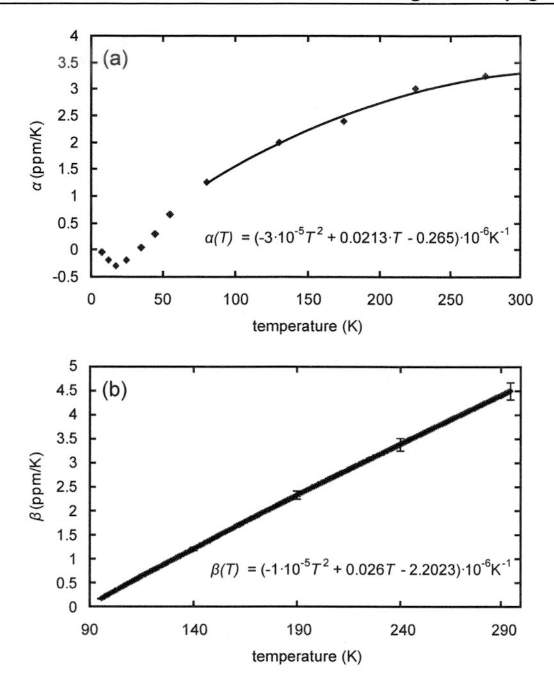

FIG. 3.15. (a) Temperature dependence of the linear expansion coefficient $\alpha(T)$ in borosilicate glass (from [13]). The solid line represents a quadratic fit within the relevant temperature range (see Eq. (3.20)). (b) Temperature dependence of the thermo-optic coefficient $\beta(T)$. The curve is based on the experimental data presented in FIG. 3.14 and gained under the assumption that the presented theoretical model of temperature tuning is fully applicable to microbubble resonators. A quadratic fit in the temperature range under investigation is also given (see Eq. (3.21). The errors are exemplarily indicated at elevated temperatures.

From the experimental data, a room temperature value of $\beta_{295K} = (4.48 \pm 0.19)\cdot 10^{-6}\ \mathrm{K}^{-1}$ was determined for the thermo-optic coefficient in borosilicate glass. In the LN temperature range, this value decreases to $\beta_{90K} = (0.05 \pm 0.01)\cdot 10^{-6}\ \mathrm{K}^{-1}$. The errors of these two values are mainly based on the fitting accuracy of the referenced $\alpha(T)$ and the errors determined by the DSRs of the recorded data sets. Lacking information about the room temperature and LN temperature values of the thermo-optic coefficients of pure borosilicate glass, these results can not be validated by actual data from the literature. However, given the very good agreement between the theoretical estimations and the measured data sets, these values seem at least quite reasonable. The fully measured temperature tuning curve for the thermo-optic coefficient can be seen in FIG. 3.15b. This curve was again fitted by a quadratic function for the temperature range under investigation. The temperature dependence of the thermal expansion coefficient is thus approximated by

$$\beta(T) = (-1\cdot 10^{-5}\cdot T^2 + 0.026\cdot T - 0.2023)\cdot 10^{-6}\,\mathrm{K}^{-1}. \qquad (3.21)$$

The final results of the pressure tuning experiments under room temperature and LN temperature conditions are presented in FIG. 3.16. The room temperature tuning curve is in perfect agreement with the given theoretical estimations based on the geometric parameter χ and by considering Eq. (3.17). The experimentally measured pressure tuning slope for the examined microbubble is nearly –15.1 GHz/bar as the theory predicts. This again verifies the applicability of the theoretical model to describe the pressure tuning abilities of microbubbles as introduced in Section 3.5.

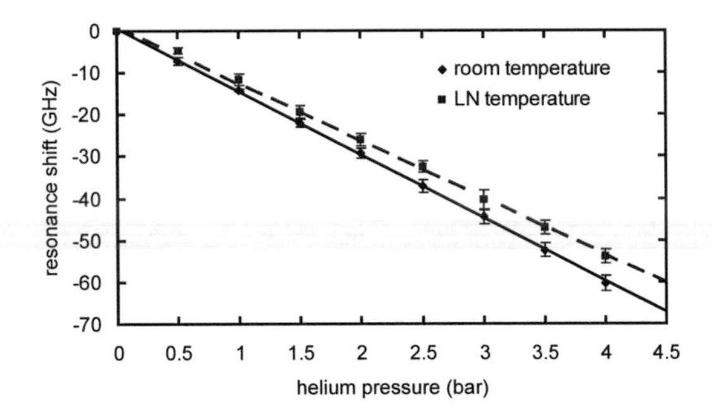

FIG. 3.16. Pressure tuning curves for a microbubble at room temperature and LN temperature conditions. The observed room temperature tuning rate for this microbubble corresponds well with the estimated value from its geometric parameter χ.

By acquiring the two curves presented in FIG. 3.16, the different measurements were repeated several times and the resulting errors were calculated from these data sets. These errors are presented as bars at their corresponding data points. Within these error bars, the two different curves for the room temperature and the LN pressure tuning show slightly different slopes. At LN temperature, the measured tuning slope seems to be a bit lower compared to the room temperature value. This behavior is in perfect agreement with the theoretical estimation about the temperature dependent pressure tuning properties. For the experimentally measured LN pressure tuning slope, a value of nearly –13.7 GHz/bar is found.

An unknown parameter in Eq. (3.16) is the temperature dependence of the thermo-optic coefficient $C(T)$ in borosilicate glass, so the presented theoretical model can be used to experimentally determine this property at elevated temperatures. For that, it is assumed that the model describes all the effects observed by pressure tuning real resonators without deviation. The given model is strictly valid only for microbubbles with perfect spherical symmetry and homogeneous wall thickness. A microbubble always exhibits residual capillary ends normally breaking the symmetry. Due to irregularities during manufacturing, also small differences in the wall thickness are likely. However, based on the good agreement of the pressure tuning results with the assumptions of the model, the thermo-optic coefficient at LN temperature can be measured by comparing the room temperature and LN temperature pressure tuning curves.

Therefore, it is assumed that the temperature dependence of the elastic shear modulus of Pyrex glass is also valid for the special borosilicate material used for producing the microbubbles. The difference between the room temperature and the LN temperature pressure tuning curves can then directly be used to estimate the cryogenic value of the elasto-optic constant C. From measurements, a decrease between the referenced room temperature tuning slope and its cryogenic counterpart of nearly 10 % is observed. This directly transfers to $C_{90K} = (3.55 \pm 0.08)\cdot 10^{-12}$ m^2/N for the elasto-optic constant at temperatures around 90 K. The given error is based on the measured statistics and the fitting accuracy of the Pyrex reference values.

3.8 Summary and Outlook

In this chapter, a new type of WGM microresonator was introduced and investigated for its tuning abilities. In contrast to some other well-known microresonator systems, microbubble resonators offer a variety of new experimental possibilities due to a free access to the inner cavity volume. In combination with simple manufacturing methods, which do not require a clean room or lithographic techniques for production, this type of microresonator is perfectly suited for all kinds of microresonator-based research and photonic applications. It is a simple but very versatile alternative to modern chip-based microfluidic devices or resonator systems.

In sensing applications, where the interaction between a WGM and a specific medium is often analyzed by observing modal shifts, these properties can be used to highly reduce the required amount of trace gas or liquid. By using a two-port microbubble, dynamic microfluidic analyses with ultra-low requirements on the detection volume are conceivable. As for most of such spherical resonator designs, in sensing applications coupling to WGMs is performed via a fiber taper from outside the cavity. By separating the coupling region from the detection volume, the fiber taper is completely unaffected by the detected media. In this way, the optical parts of the detection system are fully isolated from their functional parts and long-term stable working conditions can be achieved. For such applications, classical microspheres have been studied for a long time, but a commercial utilization could not be shown so far. Microbubble resonators could change this due to their simple manufacturing and handling characteristics.

FIG. 3.17. Microbubble with a diameter of 232 µm filled with a CdSe/ZnS core-shell quantum dot solution. The quantum dots are excited via the fiber taper.

Another aspect which makes microbubbles interesting for sensing applications, is the possibility of enhancing the interaction of the higher order WGMs with the inner media in thin-walled microstructures. In these cases, a large evanescent part of the circulating fields is exposed on the inner wall of the microbubble. This effect enhances the overlap between an inner media and the WGMs such that the detection efficiency of the system is highly increased. By that, also the effective detection volume is enhanced due to the larger evanescent fields.

Beside classical sensing applications, the involved effects can also be exploited for enhancing the interaction between the WGMs and an optically active media. By filling a microbubble with an active dye or rare earth solution, light amplification, frequency conversion, or even lasing can be observed. It depends on the specific experimental conditions how this can be achieved in detail. However, a microfluidic system which continuously provides an optical active media should be suitable for such applications. In first experiments with single-sided microbubbles, the general functionality of such filled devices could already be demonstrated [15]. In FIG. 3.17, an example is shown where CdSe/ZnS core-shell quantum dots are excited by a fiber taper.

Within this chapter, a detailed analysis of the pressure tuning abilities of microbubble resonators under room temperature and cryogenic conditions was presented. This is specifically important for applications requiring wide tunability of WGMs inside a microresonator, e.g., lasing applications where the WGM resonator is used as feedback device to select a specific frequency inside the gain region of a laser medium. Also for applications where single molecules or defect color centers with specific emission lines have to be matched with distinct WGM resonator lines, the wide-range tunability of WGM resonances is an important feature in resonator design. The figure of merit in such applications is normally defined by maintaining a small intrinsic linewidth in combination with a large possible tuning in the range of a FSR of the system. Although a variety of tuning techniques is known for different resonator designs, none of them is perfectly suited for long-range tuning inside a cryostat. This was in fact the motivation to investigate the tuning abilities of microbubble resonators and which finally led to the presented results.

It could be shown that the given pressure tuning results of single-port microbubbles produced from borosilicate capillaries are in perfect agreement with the predictions based on the elasticity theory of spherical glass shells. By applying an internal uniform pressure to a microbubble, the geometrical dimensions of the device are changed. An additional effective refractive index change is caused by induced material stress-strain components. In result, a strong red-shift of the WGMs can be observed. The frequency shift is material dependent, but the model can be highly simplified by inserting specific material constants. A direct linear relationship between the geometric parameters of a microbubble and its pressure tuning abilities is observed. Depending on the geometric parameters, for some microbubbles resonance shifts in the range of a system FSR could be successfully demonstrated.

The two presented experiments were performed under room temperature and cryogenic conditions. In both cases, the theoretical model, which bases on elasticity theory, is in very good agreement with the experimental results. For cryogenic experiments, also the temperature dependent resonance shift of the WGMs was analyzed. As microbubbles are produced under room temperature conditions, this additional resonance shift has to be fully taken into account before manufacturing frequency determined microbubbles for low-temperature applications. For considering such a temperature-driven resonance shift, a simplified model based on the opto-mechanical properties of borosilicate glass was applied. As not all required material parameters could be found in literature, the experiments were also used to estimate these missing values.

In result, it could be shown that pressure tuning in hollow microbubble resonators is an appropriate method for shifting WGM resonances both under room temperature and cryogenic conditions. By measuring the additional temperature dependent resonance shift, the pressure tuning properties of the resonator system could be fully investigated. Thus, matching the various WGM resonances of a microbubble resonator to an applied molecule or another emitter system with a specific optical transition becomes possible even under cryogenic conditions.

Beside those direct tuning experiments, the microbubble system seems also suitable for applying an active resonance stabilization scheme. As pressure tuning is purely based on the mechanical and opto-mechanical properties, the tuning process is completely reversible. This allows for fast and reliable active tuning schemes going from aerostatic to pneumatic operation. It is assumed that the achievable tuning velocity will mainly be limited from ensuring a fast and reliable in- and outlet of the tuning gas through the tapered capillary sections. This should also open the path to advanced opto-mechanical microbubble experiments where the couplings between mechanically and optically excited breathing modes are investigated.

[1] M. L. Gorodetsky and V. S. Ilchenko, "Optical microsphere resonators: optimal coupling to high-Q whispering-gallery modes," *J. Opt. Soc. Am. B*, vol. 16, no. 1, pp. 147–154, 1999.

[2] M. Sumetsky, "Whispering-gallery-bottle microcavities: the three-dimensional etalon," *Opt. Lett.*, vol. 29, no. 1, pp. 8–10, 2004.

[3] S. Berneschi, D. Farnesi, F. Cosi, G. Nunzi Conti, S. Pelli, G. C. Righini, and S. Soria, "High Q silica microbubble resonators fabricated by arc discharge," *Opt. Lett.*, vol. 36, no. 17, pp. 3521–3523, 2011.

[4] H. Li, Y. Guo, Y. Sun, K. Reddy, and X. Fan, "Analysis of single nanoparticle detection by using 3-dimensionally confined optofluidic ring resonators," *Opt. Express*, vol. 18, no. 24, pp. 25081–25088, 2010.

[5] I. M. White, H. Oveys, and X. Fan, "Liquid-core optical ring-resonator sensors," *Opt. Lett.*, vol. 31, no. 9, pp. 1319–1321, 2006.

[6] V. S. Ilchenko, P. S. Volikov, V. L. Velichansky, F. Treussart, V. Lefèvre-Seguin, J.-M. Raimond, and S. Haroche, "Strain-tunable high-Q optical microsphere resonator," *Opt. Commun.*, vol. 145, no. 1, pp. 86–90, 1998.

[7] K. Srinivasan and O. Painter, "Optical fiber taper coupling and high-resolution wavelength tuning of microdisk resonators at cryogenic temperatures," *Appl. Phys. Lett.*, vol. 90, no. 3, p. 031114, 2006.

[8] R. Henze, T. Seifert, J. M. Ward, and O. Benson, "Tuning whispering gallery modes using internal aerostatic pressure," *Opt. Lett.*, vol. 36, no. 23, pp. 4536–4538, 2011.

[9] R. Henze, J. M. Ward, and O. Benson, "Temperature independent tuning of whispering gallery modes in a cryogenic environment," *Opt. Express*, vol. 21, no. 1, pp. 675–680, 2013.

[10] M. L. Gorodetsky, A. A. Savchenkov, and V. S. Ilchenko, "Ultimate Q of optical microsphere resonators," *Opt. Lett.*, vol. 21, no. 7, pp. 453–455, 1996.

[11] S. Götzinger, S. Demmerer, O. Benson, and V. Sandoghdar, "Mapping and manipulating whispering gallery modes of a microsphere resonator with a near-field probe," *J. Microsc.*, vol. 202, no. 1, pp. 117–121, 2001.

[12] G. Righini, Y. Dumeige, P. Féron, M. Ferrari, G. Nunzi Conti, D. Ristic, and S. Soria, "Whispering gallery mode microresonators: fundamentals and applications," *Riv. del nuovo Cim.*, vol. 34, no. 7, pp. 435–486, 2011.

[13] M. Sumetsky, Y. Dulashko, and R. S. Windeler, "Optical microbubble resonator," *Opt. Lett.*, vol. 35, no. 7, pp. 898–900, 2010.

[14] A. Watkins, J. M. Ward, Y. Wu, and S. N. Chormaic, "Single-input spherical microbubble resonator," *Opt. Lett.*, vol. 36, no. 11, pp. 2113–2115, 2011.

[15] T. Seifert, "Optische Charakterisierung von Whispering Gallery Modes in Hohlkugelresonatoren," Humboldt-Universität zu Berlin, 2011.

[16] S. Timoshenko and J. N. Goodier, *Theory of Elasticity*, 2nd ed. New York: McGraw-Hill, 1951.

[17] T. Ioppolo and M. V. Ötügen, "Pressure tuning of whispering gallery mode resonators," *J. Opt. Soc. Am. B*, vol. 24, no. 10, pp. 2721–2726, 2007.

[18] F. Treussart, V. S. Ilchenko, J. F. Roch, J. Hare, V. Lefèvre-Seguin, J.-M. Raimond, and S. Haroche, "Evidence for intrinsic Kerr bistability of high-Q microsphere resonators in superfluid helium," *Eur. Phys. J. D*, vol. 1, no. 3, pp. 235–238, 1998.

[19] M. Gregor, C. Pyrlik, R. Henze, A. Wicht, A. Peters, and O. Benson, "An alignment-free fiber-coupled microsphere resonator for gas sensing applications," *Appl. Phys. Lett.*, vol. 96, no. 23, p. 231102, 2010.

[20] P. E. Barclay, C. Santori, K.-M. Fu, R. G. Beausoleil, and O. Painter, "Coherent interference effects in a nano-assembled diamond NV center cavity-QED system," *Opt. Express*, vol. 17, no. 10, pp. 8081–8097, 2009.

[21] Y.-S. Park, A. K. Cook, and H. Wang, "Cavity QED with diamond nanocrystals and silica microspheres," *Nano Lett.*, vol. 6, no. 9, pp. 2075–2079, 2006.

[22] A. Chiba and H. Fujiwara, "Resonant frequency control of a microspherical cavity by temperature adjustment," *Jpn. J. Appl. Phys.*, vol. 43, no. 9A, pp. 6138–6141, 2004.

[23] Q. Ma, T. Rossmann, and Z. Guo, "Whispering-gallery mode silica microsensors for cryogenic to room temperature measurement," *Meas. Sci. Technol.*, vol. 21, no. 2, p. 025310, 2010.

[24] S. Jacobs, "Dimensional stability of materials useful in optical engineering," *Opt. Acta (Lond).*, vol. 33, no. 1, pp. 1377–1388, 1986.

[25] S. Spinner, "Elastic moduli of glasses at elevated temperatures by a dynamic method," *J. Am. Ceram. Soc.*, vol. 39, no. 3, pp. 113–118, 1956.

Chapter 4

Chip-Based Silica Microresonators

4.1 Introduction

Whispering gallery mode (WGM) microresonators are universal systems suitable for a wide range of applications not just in optics and photonics. Furthermore, they can be used as highly versatile optically driven sensing devices. Various examples are given in the introduction to Chapter 2. For all those applications, well-defined optical properties are the basic requirement. The possible Q factors, resonance wavelengths, and free spectral ranges of microresonators are key parameters to achieve highly controlled experimental conditions. Conventional microsphere resonators can be produced with a very simple manufacturing scheme by melting standard optical glass fibers in a CO_2 laser beam or by flame heating. However, this method lacks sufficient control over the processing parameters and reliable device manufacturing is not possible. In contrast to microspheres, chip-based microresonator designs allow a highly precise fabrication relying on standard lithographic structuring methods known from the semiconductor industry. They can directly be adopted for photonics at comparable structure sizes and allow direct chip-based hybrid integration with other optical elements. This is a major advantage of integrated microphotonics compared to conventional designs consisting of large amounts of individual components. Furthermore, the mechanical stability of such systems can be highly improved with integrated photonics.

Silica is the material of choice for most applications working in the visible (VIS) and near infrared (NIR) spectral ranges. It is a stable and robust material with exceptional optical properties. As the characteristics of microresonators strongly depend on these properties, within silica, highest Q factors can be achieved. Silica can be grown or oxidized on standard silicon wafers. Silicon-based materials are the main standard of the semiconductor industry. Therefore, a large number of structuring techniques were developed and can be applied for processing.

Most microresonator applications require very wide tunability for the observed WGM resonances. There are various methods known, all having their specific advantages and disadvantages concerning applicability, achievable tuning range, reversibility, tuning speed, and resolution. In quantum optics, usually frequency gaps of more than the free spectral range (FSR) in the resonant systems have to be spanned with highest precision. Thereby, the measured Q factors should not degrade. Applicability in a cryostat is a further requirement within this field. De facto, nearly all practical applications require an appropriate resonance tuning scheme. For chip-based microresonators, also the influence of the substrate must be considered. Hence, extensive investigations on the tuning abilities of integrated systems are required.

In the following chapter, chip-based resonator structures are investigated. After a short introduction to general lithography, the fabrication process of high-Q disk and toroidal silica microresonators is presented. For tuning the WGM resonances of such chip-based microstructures, different methods and techniques are summarized. This includes the well-known temperature tuning as well as other intrinsic and extrinsic tuning methods used for specific sensing or lasing applications. The two types of resonators are then closely examined by experimental investigations. For permanently tuning disk-type microresonators, a novel post-production tuning scheme is presented [1]. The discussed technique allows a selective resizing of chip-based microresonators with high accuracy. This method is ideally suited for cryogenic applications where other tuning techniques are quite hard to implement. As direct example for such an approach, a corresponding quantum optics experiment is presented [2]. Single photon emitting defect centers in nanodiamonds are attached to toroidal microresonators for investigating the properties of the combined quantum system. The presented measurements were performed under room temperature and cryogenic conditions which show the general applicability of chip-based microresonator designs for cavity quantum electrodynamics (CQED).

4.2 Fabrication

In contrast to manually produced microresonators, i.e., conventional microsphere or microbubble WGM resonators, the highly controlled semiconductor-based processing techniques, which are applied for fabricating the chip-based resonator designs, are widely extending the range of possible applications. Well-defined structuring of multiple optical components becomes possible within a single production step. The produced devices are all located on the same microchip and can be combined with different functionalities. The used wafer-based processing techniques are scalable and allow the high-density integration of complex hybrid systems. Therefore, even for first experimental lab-on-chip designs, the direct development of chip-integrated photonic structures offer many advantages over the conventional resonator designs and setups consisting of multiple bulk optical elements.

In the following section, the basic techniques for producing chip-based microresonators are briefly discussed. Although the presented methods are specific to silica-on-silicon substrates, the general lithographic schemes and etching techniques are also applicable to a wide range of other material systems. This section gives a rough guideline through production and design. A more detailed summary about the fabrication of chip-based microresonators can be found in [3].

4.2.1 Wafer Oxidation

The basic material for the fabrication of chip-based silica microresonators are standard electronic grade silicon wafers. These wafers are originally made for the production of components from the semiconductor industry. An extremely low density of material impurities is guaranteed by the manufacturer (< 1 ppm). For photonic applications in silica, a comparable material quality is required. By thermally oxidizing silicon wafers layer by layer, a high-quality thin film of ultra-pure silica can be produced from such electronic grade materials.

Even under normal ambient conditions, a naturally appearing layer of silica can often be observed on silicon wafers. These layers have thicknesses between 0.7 nm to 3 nm and must be removed before further processing steps can be applied [3]. The oxide layer passivates the material surface and reduces the chemical reactivity.

For producing much thicker layers of silica, thermally enhanced oxidation in an oxygen enriched atmosphere has to be used. In a special furnace, the oxidation rate is increased at temperatures around 1200 °C [3]. The oxidation can be performed under dry or wet chemical processing. The dry process produces silica layers with a higher material density (2.27 g/cm³) compared to the wet chemical process (2.18 g/cm³) [4]. However, the wet chemical process is nearly 10 times faster than the dry variant. For silica layer thicknesses above 1 μm, wet chemical oxidation is commonly applied [3]. For this process, water vapor is used as wet oxidant. The involved reaction is described by

$$\mathrm{Si} + 2\mathrm{H_2O} \Rightarrow \mathrm{SiO_2} + 2\mathrm{H_2}. \tag{4.1}$$

The vapor molecules are adsorbed by the surface of the wafer. Afterwards, they diffuse into the material. At the boundary between silica and silicon, these molecules react with the silicon. Hence, the produced oxide layer is constantly growing into the material. Due to the different material densities of silicon and silica, the thickness of the wafer is slightly increased during oxidation.

For an ultimate purity of the oxidized silica, undoped silicon wafers are preferred as source materials. If only doped wafers are available, materials with segregation coefficients $k > 1$ should be used. This coefficient describes the ratio between the solubilities of the specific dopants inside silicon and silica and thus determines their preferred localization after the oxidation has finished. In $k > 1$ materials, the so-called pile-up effect ensures that most of the dopants are deposited within the substrate and thus do not influence the purity of the produced silica layer.

4.2.2 Photolithography

The first step in microresonator production is patterning structures on top of the wafer. Therefore, a light-sensitive photoresist is used to transfer a two-dimensional geometric pattern from a photomask or by a direct writing scheme onto the substrate. This pattern transfer can be done positive or negative depending on which part of the resist becomes soluble to a developer solution after exposing to ultraviolet (UV) light. If the exposed portion of the photoresist becomes soluble, then the material behaves positive. If the unexposed parts of the resist are dissolved, then the resist behaves negative. Both types of photoresist have specific advantages and disadvantages. However, for achieving the highest possible resolutions, positive resists are preferred. Different types of materials are available for using as resist. The most common negative resist is the epoxy-based polymer SU-8. It can be structured with standard mask-based processing techniques or by direct e-beam writing.

Before a photoresist can be applied to the wafer, the surface of the wafer has to be cleaned from any organic or inorganic contaminations. This is often performed by a set of consecutive wet chemical treatments. A standard for wafer cleaning is the so-called RCA (Radio Corporation of America) process. It contains several baths in organic and inorganic solvents and is performed at specific processing temperatures for each of the baths. For silicon processing, the removal of the naturally appearing silica oxide layer is included within the RCA process.

The different photoresists are normally applied by spin coating. Therefore, a wafer is mounted on its central axis and spun with up to 5000 revolutions per second. By dispensing a viscose solution of resist onto the top of the wafer, a highly uniform layer can be produced. The layer thickness is controlled by the concentration of solvent within the photoresist and by the rotational speed of the used spin coater. For most processing techniques, a resist layer thickness between 0.5 µm and 3 µm is required. After spin coating, the residual solvent is driven off by a thermal pre-bake typically performed at temperatures around 100 °C.

For the production of disk-type microresonators, the basic structuring pattern consists of multiple circular disks. The diameters of these disks vary between a few and up to several hundreds of micrometers depending on the desired application. If a mask-based process is used, a photomask has to be written in advance by e-beam or laser beam writing. The masks are aligned in parallel to the resist surface and intense UV light is used to chemically modify the different areas. A photomask can easily be used several thousand times and thus mass production of photonic elements becomes possible. However, a more flexible lithography scheme is often desired for simple design testing.

In this case, direct structuring of the photoresist via e-beam direct writing lithography can be applied. This technique allows the highest accuracy for defining the structures and enables simple modifications of the produced designs. Due to the intrinsically high resolution of the structuring process, this type of lithography is time consuming and expensive. Hence, this technique is merely for preliminary design studies and other techniques have to be applied for later mass production. After all exposure processes are finished, a post-exposure bake has to be performed to enhance the induced chemical reactions inside the photoresist.

With both structuring methods, the developed areas are subject to permanent localized structural changes. Depending on the selected type of photoresist, these areas become or stay soluble for specific developer solutions. For that, the full wafer or microchip is submerged into an appropriate developer bath and then cleaned with deionized water. The lithography process is finished by solidifying the resist in a final hard-bake. This step increases the durability of the structured areas and high resistivity against the subsequent etching processes is achieved. A standard hard-bake is often performed at temperatures of up to 200 °C for more than 30 minutes. At these temperatures, the resist partly reflows and increases the definition of the structural patterning.

4.2.3 Etching

Within the photolithographic processes, well-defined structures are patterned into the previously deposited photoresist. These patterns protect the underlying material against the further processing steps. The unprotected areas are fully exposed to the external environment and can be removed by various etching methods. In general, two different etching processes are distinguished.

One method is based on wet chemical etching. By emerging the wafer or microchip into an etchant bath, highly isotropic material removal can be achieved. This is due to chemical dissolving processes which homogenously act on all the exposed surfaces. However, by etching crystalline materials, highly anisotropic etching behavior can be observed. The various crystal facets often interact differently with the etchant and the resulting etching rates may substantially differ between these surfaces. For etching silica-based materials, a buffered hydrofluoric acid (BHF) solution is commonly used. This aqueous solution of hydrogen fluoride shows good separability against silicon and most standard photoresists. A typical buffering agent is ammonium fluoride (NH_4F). By adding a buffer to the solution, the controllability of the etching process is highly enhanced [5]. The silica is removed in a dissolution process described by

$$SiO_2 + 4HF \Rightarrow SiF_4 + 2H_2O\,. \tag{4.2}$$

The dissolving rate is temperature dependent. By heating the etchant bath, the observed etching rate can be increased. For a better homogeneity of the process, the bath should also be stirred during etching. This actively removes the reaction products from the wafer surface and replaces them with new unreacted reactants.

The second method uses reactive ions from a gas phase to remove the surface material layer by layer. This so-called dry etching process is performed in an etching chamber where reactive gas plasma is constantly generated by microwave or other excitation. Depending on the kinetic energy of the produced ions, dry-etching processes can be divided in three different regimes. These are chemical plasma etching (PE), reactive ion etching (RIE), and physical ion beam etching (IBE) [6]. In FIG. 4.1, the different types of dry etching are illustrated. The three micrographs show the typical silica-on-silicon structuring results observed for these processes.

In the PE regime, the kinetic energy distribution of the ions is isotropic. The velocity distribution corresponds to a simple Maxwell-Boltzmann distribution. In this case, the reaction of the ions is purely chemical and thus the performance of the etching process highly depends on the type of applied ions. The etchants react with molecules on the surface of the wafer. These reactions produce highly volatile reaction products which have to be constantly removed from the etching chamber. The PE process is directly comparable to a wet chemical etching approach.

In the IBE regime, the ions are accelerated to the wafer surface by external electric fields. Due to the high kinetic energies of the incoming ions, the material removal is mainly carried out by unselective sputtering processes. As for this type of dry-etching only the energy of the impacting ions is important, the process is not particularly dependent on a specific material and nearly all types of ions can be used for the bombardment. A major advantage of this method is the high level of directional control. By tilting the wafer, even slanted walls can be etched.

The RIE process is directly in between the two previous regimes. It is even-handedly performed in a combination of physical and chemical reactions. In addition to the acceleration of reactants, the incoming ions can chemically bind to surface molecules or atoms. Thus, the sputtering efficiency is enhanced and highest etching rates can be observed for this process. With this technique, the etching mainly acts perpendicular to the wafer surface and thus deep trenches with straight sidewalls can be produced with this method.

The three different etching regimes can be further categorized by their processing speed, material selectivity, and the achievable degree of etching isotropy. With ion bombardment, the surface roughness of the produced structures is commonly much higher as compared to a purely chemical etching approach. Thus, for producing microresonators with highest surface quality and Q factors, a chemical PE process is the method of choice. For non-crystalline materials, highly isotropic etching behavior can be observed with this method. The achievable degree of isotropy is comparable to wet-chemical etching, but the results are often more homogeneous along the surface. The degree of isotropy, resulting etching rates, and selectivity against other materials can be improved by adding some specific supplemental gases to the etching chamber. Depending on the etching process, argon (Ar), oxygen (O_2), or nitrous oxide (N_2O) can be applied for processing [4]. Furthermore, the etching process can be controlled by regulating the physical conditions inside the chamber. Varying the different gas flows, the microwave power for ion creation, or the partial pressures of the involved gases allows to influence the etching process over a wide range of parameters [6].

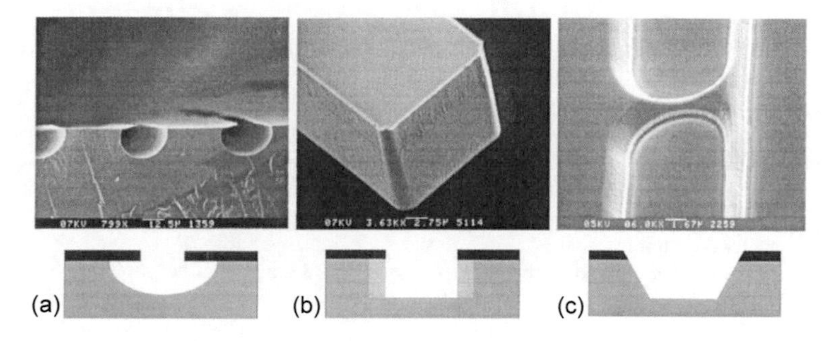

FIG. 4.1. Electron micrographs of typical etching results achieved with three different types of plasma etching at silica-on-silicon substrates (black-on-grey) (from [6]). (a) Plasma etching (PE). (b) Reactive ion etching (RIE). (c) Ion beam etching (IBE).

For the chemical PE of silicon, different gases can be used. Most common are sulfur hexafluoride (SF_6) and xenon difluoride (XeF_2). In both cases, fluorine ions are the reactive species which is responsible for the material removal. In contrast to a wet chemical etching approach with BHF, fluoride-assisted PE allows much higher etching rates on silicon, but can also be used for structuring silica patterns. For XeF_2, the etching process of silicon is described by

$$2XeF_2 + Si \Rightarrow SiF_4 + 2Xe. \tag{4.3}$$

One product of this reaction is silicon tetrafluoride (SiF_4). This highly volatile gas can easily be removed from the chamber. In a SF_6-based dry etching plasma, a similar chemical process is observed [7]. The initial sulfur molecule is reduced by multiple fluoride-assisted etching steps before SiF_4 as final reaction product is produced [8].

4.2.4 Post-Processing

By performing the etching process, the previously structured resist pattern is transferred into the substrate. In case of disk-type microresonators, circular silica disks are created.

The resulting disks are still covered by photoresist. For removing these residua, a liquid resist stripper can be used. It modifies the chemical binding between the photoresist and its underlying surface such that the resist can be simply washed off. Another option to remove residual photoresist is by oxidizing it in oxygen plasma. The corresponding process is called "ashing". It can be performed in a special ashing chamber or by directly using the dry etching chamber after the standard etching processes have been finished. For finalizing the post-processing of wafers or chips, they are again cleaned in an RCA process.

The achievable surface quality of the etched structures strongly depends on the used processing parameters. By modifying the conditions inside the chamber or by changing the applied etching method, the result of an etching process is widely controllable. A final hard-bake of the etched structures can also be applied to further improve the result. Furthermore, it is often possible to enhance the resulting surface quality by performing a final dip-etch in a highly concentrated wet chemical etching solution. With such a technique, surface roughnesses in the nanometer range can be realized. However, the dip-etching method only allows polishing of small-scale roughness as it is typically observed with dry etching processes. If instead the lithographic process itself or the used photomask provide insufficient resolution, then the roughness of the structured patterns is limited by this effect and can not be decreased by dip-etching. In FIG. 4.2, an example of such an unpolishable large-scale roughness is presented.

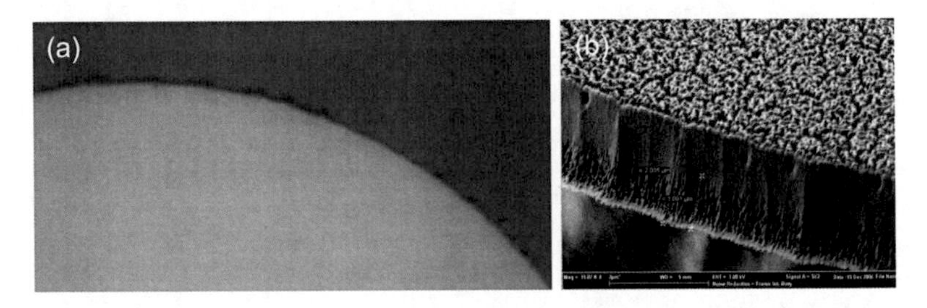

FIG. 4.2. (a) Microscopic photograph of a microdisk with a diameter of 68 μm. The disk was wet chemically etched by a BHF solution. The cogwheel-like features on the edges of the structure are due to an insufficient resolution of the lithographic mask. (b) Electron micrograph of the same resonator structure. The cogs on the edge are clearly visible.

Another possibility for decreasing the final surface roughness is by performing an additional thermal treatment. In some organic materials, hard-bake like processing can be applied to temper and thermally reflow the etched structures. In these cases, surface tension is used for smoothing by heating the material over its specific glass transition temperature. This surface tensioning effect can also be used for silica by heating the structures with a CO_2 laser. Although silicon oxide is highly absorbing at this specific laser wavelength (λ = 10.6 µm), silicon is here more or less transparent. Furthermore, the silicon stems underneath the disks act as heat sinks due to their high thermal conductivity. Hence, by applying a CO_2 laser beam to the structures, mainly the silica annulus around the stem is locally heated. By carefully adjusting the applied laser power, the heating process can be widely controlled and even melting and reshaping of the structures becomes possible. This surface tension induced reflow process is already known from conventional microsphere manufacturing. Here, this effect directly leads to the on-chip production of toroidal microresonators [9]. Within these structures, highest finesses can be realized due to the exceptional quality of the resulting surfaces. Furthermore, due to the applied thermal treatment, a rearrangement of the amorphous silica can be observed by an increased density of the reflowed material [10].

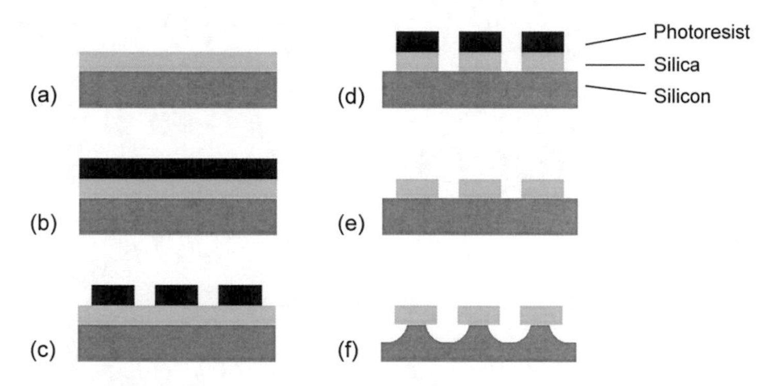

FIG. 4.3. Schematic of the production process for chip-based disk-type silica microresonators. (a) Silicon substrate with silica layer on top. (b) Application of photoresist. (c) The resist layer is patterned with circular disks by mask-based photolithography. (d) The pattern is transferred to the silica by isotropic wet chemical etching. (e) The residual photoresist is lifted off from the structures. (f) The silicon underneath the disk structures is selectively removed by an isotropic PE process. The structures are finalized by a short dip-etch in a concentrated BHF solution.

The separation of single microchips can be performed with different methods. It can be carried out by diamond scribing and adjacent breaking, by sawing, or by laser cutting. For dicing a wafer or parts of a wafer, they are mounted on a special dicing tape which allows simple handling throughout the full process. The designs for microchips often include so-called dicing streets. Along these lines, the cutting process can be performed without affecting the desired functional structures. For silica-on-silicon structures, all available dicing processes can be applied. However, sawing often causes high-frequent mechanical vibrations traveling along the surface of the wafer. In such a process, without special precautions, fragile microstructures, e.g., free standing silica disks or toroids, can easily break off. For a complete underetching of such structures, it may be required to dice the wafer before the final underetching is performed.

4.2.5 Results

All silica-based microresonators shown within this thesis were produced with methods presented in the last section. The structures were processed at the cleanroom facilities of the Paul-Drude-Institut für Festkörperelektronik located in Berlin. Later works were performed by the Ferdinand-Braun-Institut, Leibnitz-Institut für Höchstfrequenztechnik in Berlin-Adlershof.

The applied fabrication process for silica-based microresonators is schematically shown in FIG. 4.3. Thereby, only major processing steps are presented. An actual processing scheme may involve additional tempering and baking steps as well as some additional cleaning procedures. With the given processes chain, the highly reliable production of silica-based disk and toroidal microresonators is demonstrated. The production process was performed at various structure sizes and for different silica layer thicknesses.

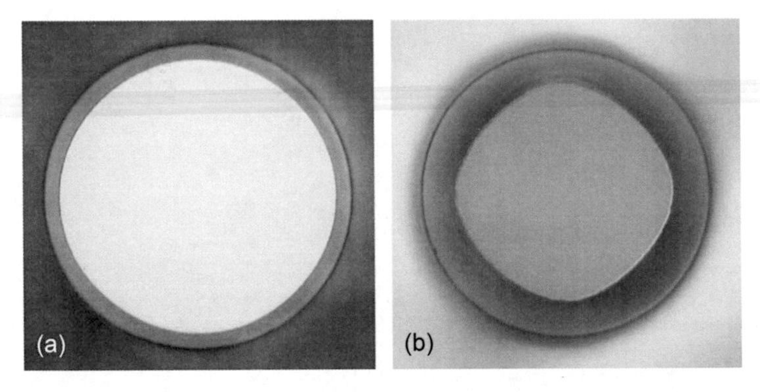

FIG. 4.4. Microscopic photographs of silica microdisks with a diameter of 120 μm. The disks were underetched with different length. The inner circle shows the silicon stem with a silica layer on top, the annulus around the stem is made of pure silica. Outside these regions, the silicon substrate can be seen from top. (a) The annulus has a width of 3 μm. (b) The annulus has a width of 20 μm.

Microdisks with diameters between a few and several hundreds of micrometers were produced. Thereby, the standard thickness of the silica layer was 2 μm. With this thickness, single-mode behavior is expected for disk-type resonators working within the telecommunication range. In this type of microresonators, various underetching lengths can be realized without affecting the optical properties of the WGMs. For the applied PE process, an aspect ratio of 1:2 was measured between the horizontal and vertical etching rates. Thus, for disk-type microresonators, the underetching parameter can be used to control the distance between the lower side of the silica layer and the base of the substrate. With a larger underetching, the supporting stems become higher and external optical coupling to the resonators is simplified for the price of slightly decreased mechanical stability. In FIG. 4.4, two disk-type microresonators are shown with different underetchings. On these two structures, nearly identical optical properties were observed.

FIG. 4.5. Silica microtoroid with a diameter of 90 µm seen from top by an optical microscope. The disk preform was underetched by 10 µm. On the right side of the picture a breaking edge of the silicon substrate can be seen. The rim of the toroid reaches over the substrate. Hence, optical coupling to the resonator can be performed with a prism or other coupling devices.

For the production of toroidal microresonators, silica disks are thermally threaded by a CO_2 laser beam such that the edges of the disks are reflowed from surface tension. A more detailed analysis of the underlying processes can be found in [11]. A complete description of the used setup is presented in [3]. The initiated reflow process shows a self-terminating behavior when all excess material is completely reshaped. Thus, for toroidal microresonators, the length of underetching is a geometric design parameter as it defines the minor diameter of the produced torus structure. For most applications, a total underetching of 10 µm is sufficient.

4.3 Resonance Tuning Methods

The high Q factors achievable with WGM microresonators make them very versatile systems for applications where small linewidths and long photon storage times are crucial parameters. For sensing and signal filtering, these well-defined properties are key issues. On the other hand, these characteristics become critical when it comes to the task of matching intrinsic resonance lines with externally defined systems and devices. Hence, when working with optical WGM microresonators, it is often required to tune a single cavity resonance to a specific frequency reference. For using such type of resonators as feedback elements in laser applications, the cavity resonances have to be matched and permanently fixed to external laser emission lines [12]. For high-Q microresonators working in the optical range, the FSR of such systems can be in the order of some nanometers and setting a specific resonance condition becomes difficult in these cases. It may be possible to achieve frequency matching by changing external parameters, e.g., by detuning the laser emission to go into resonance with a cavity.

However, in cases where such a scheme is not feasible, a reliable fine-tuning method for an internal shifting of observed WGM resonances is urgently required. Especially for controlled quantum optics experiments dealing with single quantum emitters, such as nitrogen vacancy (N-V) defect centers in diamond (see Chapter 4.5), an intrinsic resonator tuning method is necessary due to the limited external tuning abilities of N-V defect centers and the wide frequency spread of their zero-phonon emission lines. For such experiments, the resonance frequency of the cavity has to be perfectly matched with the transition line of the attached photon emitter. Hence, resonator tunability is a crucial parameter. Although these coupled systems can be experimentally analyzed at room temperature, cryogenic conditions are much more preferred for such applications. This is due to the suppressed interaction of the quantum emitters with the environment, e.g., visible in a reduced decoherence rate. Attempts to couple the narrow zero-phonon emission line of N-V defect centers in diamond nanocrystals (DNCs) to high-Q WGMs have been reported by several groups [2], [13], [14], but the required reliable tuning method under cryogenic conditions has not yet been demonstrated.

In the following section, different tuning methods for matching WGM resonances with externally defined systems are discussed. This covers well-known general WGM tuning techniques suitable for room temperature and cryogenic applications as well as specific methods for tuning the zero-phonon emission lines of N-V defect centers.

4.3.1 Direct Tuning

A directly frequency matched production of chip-based resonators is not possible in the case of high-Q microcavities. Assuming an operational wavelength of 1550 nm and an allowed maximum resonance deviation of ±0.1 nm (±11 GHz) it follows that for silica systems the minimum production precision compared to the overall size has to be in the order of ±0.006 % [1]. For most microresonator production schemes, this high level of accuracy is hard to achieve. Even with standard up-to-date lithographic techniques, designed for a highly reliable electronic microchip production in fabrication plants, such elaborated production precision is not manageable. The lithographic structuring and subsequent etching processes generate random deviations which dominate the achievable precision for defining optical structures. Furthermore, the actual resonance frequency of a WGM microresonator also depends on other physical parameters, e.g., environmental temperature and humidity. For practical applications, additional active tuning schemes must be applied in any case. Even for under some specific conditions perfectly matched microresonators, an active stabilization scheme to compensate for environmentally driven resonance fluctuations is always required.

For coupling a narrow-band light source, e.g., a specific laser or spectral emission line, to a resonator, well-defined post-production tuning is absolutely necessary. The most common tuning methods for microresonators can be distinguished into two types. These are namely the so-called intrinsic and extrinsic tuning approaches. While the intrinsic methods work on the physical properties of the microresonators, the extrinsic methods leave the cavity resonances unaffected and frequency matching is performed by modifying the properties of an external emitter or device. Most of these methods were originally developed with specific applications in mind, but they can often be used for other purposes. It depends on the actual resonator material, the experimental conditions, and the required tuning ranges which of those techniques can be applied to a specific resonator design.

4.3.2 Intrinsic Tuning

For all WGM microresonators, their specific resonance frequencies are determined by two parameters. These are the effective radius r_{eff} and the effective refractive index n_{eff} of the modes inside such cavities. Both parameters can be exploited for tuning the resonance frequency of a WGM with highest precision (see also Eq. (3.7)).

The effective refractive index n_{eff} strongly depends on the spatial distribution of the modes, and the refractive index contrast between the resonator material and the local environment. For WGMs showing the largest evanescent fields at the outside of the resonator, e.g., for the fundamental modes, the effects which are induced by changing the external environment are much higher as compared to the influence on higher order modes with their smaller evanescent fields. These modes are much stronger bound to the material and thus less sensitive to external changes. For lower-order modes, a resonance shift can be realized by simply changing the dielectric properties of the environment. With such a technique, frequency detuning of up to several GHz can be achieved [15]. If the refractive index of the environment is changed, then n_{eff} follows accordingly. When high-Q WGM resonances are observed, this effect can also be used for implementing environmental sensing capabilities [16]. In such applications, the actual resonance frequency of a cavity is continuously measured and changes in the environment are detected by measuring the induced resonance shifts.

Another possibility to change n_{eff} is by directly modifying the dielectric properties of the resonator material. This can be done by applying external currents [17], [18] or electric fields [19] to the resonator system. However, these methods are mainly limited to specific sets of conductive or non-isolating resonator materials. In some materials, also external electric fields can be applied to use field induced stresses for tuning [20]. With these material-based approaches, maximum resonance shifts in the order of some few GHz are achievable [21].

However, the far most common technique for tuning WGM resonators is by changing their actual physical dimensions, i.e., by controlling the effective radius r_{eff} of a mode. This can be done in several ways, often simply the temperature of the systems is modified and thus thermo-optic and thermo-mechanic effects are used for tuning the microresonators [22], [23]. This temperature tuning method can be applied to nearly all types of resonator materials, but it is practically limited by the accessible temperature ranges. In silica, this method allows for a differential shift rate of just a few GHz/K at room temperature. It is therefore practically limited to tuning ranges of ±50 GHz [15]. However, in a cryostat, thermal tuning is not directly applicable and thus this method is not appropriate for cryogenic applications [24], [25].

Another option to change the dimensions of a resonator is by carefully removing [26] or adding material to the surface of the resonator. The latter is relatively difficult to achieve with standard chemical deposition techniques, but in a cryostat controlled nitrogen gas adsorption can be applied [27]. By using a special cycling process, this method even allows for a reversible up and down shifting of resonances. The method gives wide tuning ranges (some nanometers for GaAs/AlGaAs microdisks), but strong Q factor degradation is observed. This is probably due to an enhanced sub-wavelength optical scattering which is induced by the adsorbed nitrogen film [27].

For specific types of resonators, further tuning methods are available. In microspheres, compression or stretching via piezoelectric actuators can be used to apply mechanical strain to the resonators [15], [28]. Also with this technique, the size of the resonator is mechanically changed. A similar tuning mechanism is also possible for the so-called bottleneck microresonators [29]. These are highly elongated structures in which strain tuning can be applied by stretching the cavities along their long axis.

It should be noted that most of the presented intrinsic tuning methods simultaneously act on both of the parameters n_{eff} and r_{eff}. This is due to the intrinsic correlation of the different modal properties. Therefore, it must be considered when an estimation of the achievable resonance shifts for such methods is required.

4.3.3 Extrinsic Tuning

In some applications, it is also possible to stay with a specific WGM resonance. Tuning is then performed on the frequency of light which is coupled into the resonator. In the simplest case, this kind of external tuning can be done 'passively' by using a suitable broadband source. In such a configuration, only the portion of light which matches with the internal resonance frequency of the cavity is coupled into the resonator. The cavity selects the right resonance frequency on its own. This principle can be applied for resonator stabilized laser systems when the external gain medium exhibits a sufficiently large spectral bandwidth. By applying a large amount of quantum dots or molecules, all showing narrow emission lines but emitting at slightly different wavelength, lasing can be achieved at the resonance frequency of the resonator [30]. This condition also holds for experiments or applications where just a few emitters or even single molecules are attached. If the spectral properties of the applied emitters are broad enough and at least partly overlapping with one of the resonances inside the cavity, passive mode-locking and an indirect tuning of laser emission can be achieved [31]. Hence, in applications where it is not required to use specific pre-defined wavelengths, such kind of passive tuning can be applied.

If such a passive tuning is not appropriate, active tuning can also be applied. By using an external laser source with a small linewidth, it may be possible to selectively excite specific cavity resonances by directly tuning the laser. Such type of frequency tuning is achievable for many laser systems by various methods. With external cavity diode laser systems, e.g., the New Focus Velocity Series, effective tuning ranges of up to a few ten nanometers and narrow linewidth emission of a few 100 kHz can be realized (see also Appendix B).

In applications where single narrow-linewidth photon emitters, as molecules, quantum dots, or N-V defect centers in DNCs, are attached to the resonant system, other tuning techniques have to be applied. In such cases, the internal electronic resonances of the emitters can be shifted by using the Stark effect [32], [33]. By applying an external electric field to the emitters, their corresponding energy levels can be shifted and a detuning of specific emission lines is possible. In N-V defect centers, this effect is normally limited to a few 100 MHz. However, by applying multi-axis electric fields to the emitters, the tuning capability of their zero-phonon transition lines can be enhanced to over 10 GHz [33].

4.4 Tuning by Selective Etching

The tuning methods described in the previous section mainly provide narrow tuning ranges over a few GHz. Moreover, they often depend on additional error-prone external control elements. In case of small resonators with just a few micrometers in diameter, the FSR of such systems can be quite large and span over several 100 GHz. Depending on the spectral range, in the optical regime this span corresponds to several nanometers in wavelength. For single mode microresonators, these are resonators where only the fundamental mode of propagation can be excited, matching to external reference lines can be impossible due to a large resonance gap. Another application where the tuning of WGM resonators is required are so-called photonic molecules [34]. These are structures where several resonators with identical or slightly different optical properties are combined to achieve multipath resonances. These combined microsystems are hard to control with the presented standard tuning techniques. Photonic molecules are often build from resonators which before had to be found in a very time consuming manual pre-selection process essentially by chance [35].

In such cases, permanent tuning techniques which allow a direct setting of individual resonance frequencies for the resonators seem more appropriate. As a direct structuring and manufacturing process is too critical in high-Q resonator systems, an alternative method for controlled post-production tuning is highly favorable. A possible approach is based on a small series of additional etching steps to selectively resize the physical dimensions of the microresonators. For silica microspheres, such a tuning technique was just recently presented [36]. In a microfluidic channel, different microspheres were processed in parallel by wet chemical etching with BHF. With this approach, tuning ranges of up to 430 GHz were successfully demonstrated. A degradation of the intrinsic Q factor was not observed for this process.

In this section, the etching method is applied to intrinsically tune the WGM resonances of disk-type silica-based microresonators. In contrast to ordinary microspheres, these chip-based optical cavities are produced in a highly controlled lithographic structuring process. This allows the design of fully integrated optical elements also including large arrays of microresonators. Direct interconnections between individual resonators are feasible by incorporating appropriate waveguide elements to the design. The chip-based resonators were structured with highest precision and quantitative measurements were performed to determine deviations between the produced cavities. The presented tuning method was developed to allow a fast and reliable permanent post-production tuning of microresonators in a standard laboratory environment.

The full method was developed in close cooperation with the Ferdinand-Braun-Institut, Leibniz-Institut für Höchstfrequenztechnik (FBH). The analysis of this method resulted in a publication [1].

4.4.1 Experimental Methods

Silica-based disk-type microresonators were produced with the lithographic techniques described in Section 4.2. Individual resonators with nominal diameters between 40 μm and 60 μm were structured on single microchips (1 mm x 1.5 mm). However, for the presented measurements merely resonators with a nominal diameter of 50 μm were used. The silica oxide layer had a thickness of 2 μm and was produced by thermally oxidizing silicon in a wet chemical process. All chips were processed as parts of larger wafer samples and diced out after the structuring processes were finished. As the microresonators on such chips all originate to the same wafer region, the applied processing conditions are highly comparable. The different resonators should show identical optical properties. Thus, possible deviations among the various samples can be interpreted as a measure for the reliability of the production process.

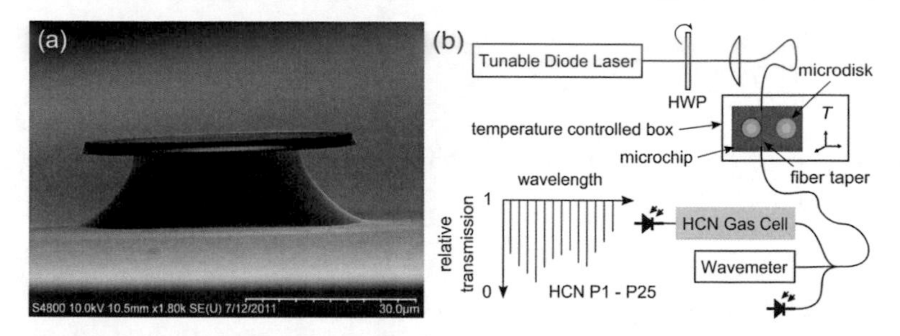

FIG. 4.6. (a) Scanning electron micrograph of a typical silica microdisk with a radius of 25 μm and a silica layer thickness of 2 μm. (b) Experimental setup for measuring WGM resonances in microdisks. The resonators are evanescently coupled by a fiber taper. For an absolute frequency reference, the wavelength of the tunable laser is continuously measured by a wavemeter. The polarization is controlled by a rotatable half-wave plate (HWP). The schematic also includes an illustration of the reference spectrum of a hydrogen cyanide gas (HCN) cell. The transmission decreases when a molecular resonance is excited by the laser.

A scanning electron micrograph of a disk-type microresonator is shown in FIG. 4.6a. The underetching of the structure was performed by anisotropic SF_6 plasma etching. For all produced microdisks, underetchings of 10 μm were performed. This results in a separation between silica and substrate of nearly 20 μm. By assuming an effective fundamental mode area of less than 5 μm^2, this separation is sufficient to ensure a negligible interaction of the WGM with the substrate or the stem of a resonator.

For analyzing the modal structure of the disk-type microresonators, a tunable external cavity diode laser (New Focus Velocity Series) in the telecommunication wavelength range between 1500 nm and 1600 nm was used. The laser combines a small linewidth (300 kHz) and a large mode-hop-free scanning range (up to 30 GHz). The laser also allows a completely mode-hop-free wavelength scan over the full wavelength range.

The absolute frequency of a disk resonance was measured by a wavemeter (Burleigh WA-1500) with an accuracy of ±0.2 ppm. A hydrogen cyanide (HCN) reference cell was used to give an absolute frequency reference for specific telecommunication C-band resonance lines. This allows a possible matching between tuned microresonator resonances and well-defined frequency reference standards.

The laser light is coupled into the resonator by the evanescent fields of a tapered optical fiber. Thereby, the fiber taper had a diameter of around 3 μm. For this size, the phase matching condition between the fiber-guided fields and the internal field of the fundamental mode in the resonator is nearly fulfilled. The position of the fiber taper was precisely controlled by a 3D piezo stage. This allows stable coupling under contact and non-contact conditions. A schematic of the full setup can be found in FIG. 4.6b.

A zero-order half-wave plate is used to rotate the polarization state of the laser beam before coupling into the fiber. The tapered region does not change the state of polarization. As this state is presumed over short distances even in non-polarization maintaining fibers, a controlled excitation of individual TE and TM modes in the resonators was achievable by simply turning the wave-plate and searching for a specific resonance frequency.

As typical for such WGM resonance laser probing experiments, the laser frequency is continuously scanned over a specific wavelength range. The distinct resonances of the cavity can then be observed by analyzing Lorentzian shaped dips in the transmitted laser power. The measurements were performed for a central wavelength of 1560 nm. After fine-tuning the laser emission to the maximum of a resonance, the wavemeter was used for a high-precision measurement of the absolute resonance frequency.

4.4.2 Characterization

In the following section, the most important optical properties for characterizing the produced disk-type silica microresonators are presented. These are the modal structure and the FSR of the resonators, the achievable Q factors, the reproducibility of a specific resonance frequency, and the thermal tuning behavior of the chip-based microdisks.

Modal Structure and FSR

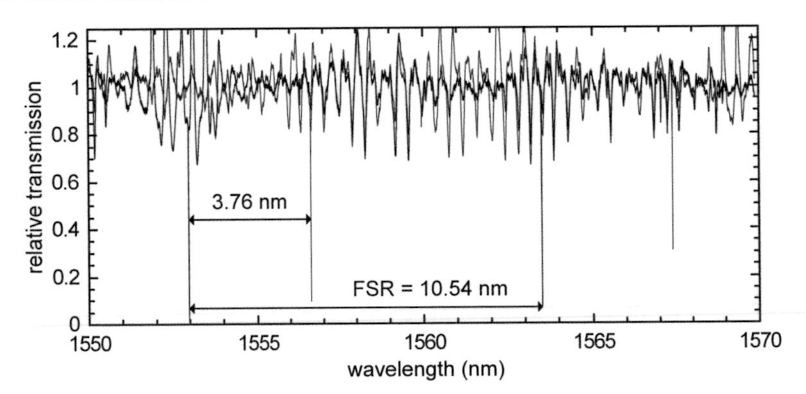

FIG. 4.7. Typical mode structure of a 50 µm disk-type silica microresonator. For wavelengths around 1560 nm, mainly the two fundamental polarization modes with a TE/TM mode spacing of 3.67 nm are observable (black and red curves). The FSR of 10.54 nm is clearly visible. It corresponds to the geometrically estimated value of (10.6 ± 0.2) nm. The sinusoidal signal around unity transmission is due to mode beating within the fiber taper. It is further amplified by the measurement process which is relative to the uncoupled taper transmission.

With the described experimental setup, it was possible to measure the initial optical properties of the resonators with highest precision. By scanning the laser wavelength over a range of several nanometers, a clearly visible mode structure could be observed (see FIG. 4.7). The determined FSR of 10.54 nm is in full agreement with the calculated FSR of (10.6 ± 0.1) nm for microdisks with nominal diameters of 50 μm. At wavelengths around 1560 nm, a medium FSR of (10.5 ± 0.3) nm was measured for the different resonators on a single microchip. Within the wavelength range of interest, the resonators show single mode behavior and mainly the fundamental TE and TM modes of the different azimuthal orders are visible. The observed single mode behavior highly simplifies the identification of individual modes. Even after extended etching steps, a direct reidentification of specific WGMs is possible as the maximum etching rate is well-known and thus an accidental over-etching of azimuthal mode orders is unlikely.

Q Factor

For the produced microstructures, initial Q factors in the 10^6 range with linewidths around 200 MHz could be measured. An example of such a resonance is presented in FIG. 4.8. The observed Q factors are comparable to the results measured by other groups and achieved in directly etched chip-based wedge-type silica microresonators with similar size [26]. Higher Q factors can be achieved within resonator designs finished by a surface tension induced thermal reflow process, e.g., for CO_2 laser treated microtoroids or microspheres. However, these designs are not compatible with standard clean room processing and require an individual thermal treatment. Another option for achieving higher Q factors is to reduce the radiation losses by using resonator designs with significantly larger effective radii [26]. Here, a precise manufacturing process is even more challenging and a reliable tuning method has still to be found.

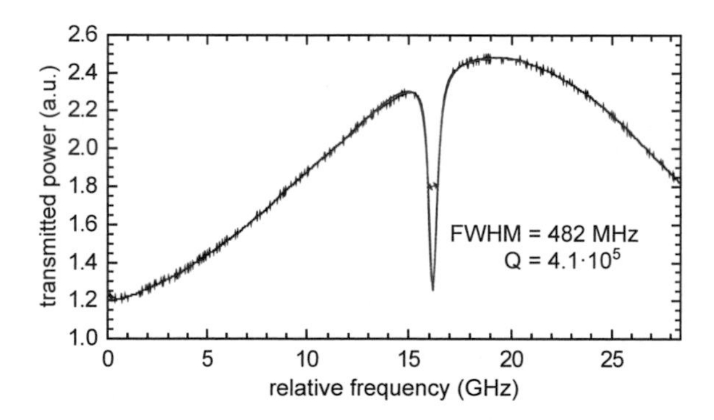

FIG. 4.8. High-resolution scan of a typical cavity resonance. The frequency resolved taper transmission of an evanescently coupled fiber taper was measured for laser wavelengths around 1560 nm. The measured Q factors exceeded 10^6 at maximum with linewidths below 200 MHz.

Resonance Frequency

For probing the reliability of the microresonator production process, a measurement of the variance in the absolute frequency of the resonances was performed. Two adjacent WGMs were searched around a central wavelength of 1560 nm. Therefore, the probe laser was scanned over a wavelength range of ±11 nm to ensure measuring the two resonances defining the FSR of the system. For measuring the precision of the whole production process, the minimum offset from the central wavelength was determined for these resonances. This offset measurement gives direct access to the degree of controllability in setting a specific resonator size during production and defines the amount of tunability which is required for achieving an arbitrary resonance frequency in later applications.

The FSR of a system is a good figure of merit to optically measure the absolute size of a microresonator. The pure resonance frequency instead requires additional knowledge about the corresponding azimuthal mode number of the WGM. When merely single resonances are measured, observed deviations between different microresonators can also correspond to WGMs with slightly different orders. While for small differences in the size of the resonators (up to some nm) the resonance frequency of the same mode order is measured, in case of larger size differences (up to some 100 nm) different mode orders may be involved. However, for the analyzed system, the modal properties of the WGMs are not much influenced by the actual size of individual resonators and thus the condition for small size differences can be assumed. Furthermore, for coupling to external frequency references, often only the absolute frequency of a resonance is important. The exact mode number of an observed WGM is practically without interest for later applications.

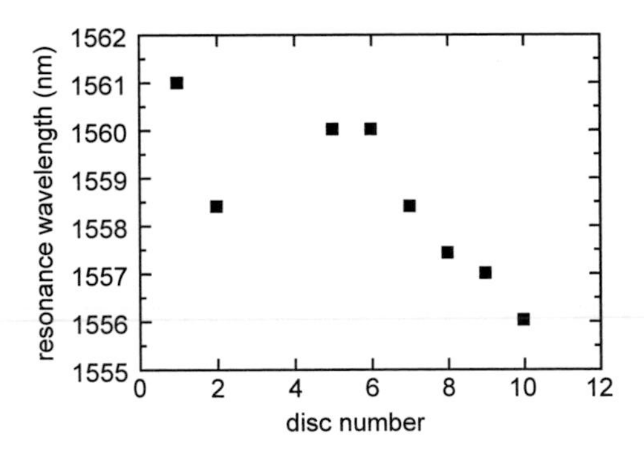

FIG. 4.9. Measured resonance frequencies of several silica disk-type WGM microresonators all produced in a common etching process. The scattering of these values is well within one FSR of the system. The individual errors per disc are below the data point size due to the precise absolute measurement of the resonance frequencies.

The result of such comparative measurements is presented in FIG. 4.9. The determined medium FSR of (10.5 ± 0.3) nm corresponds well with the theoretically estimated FSR of (10.6 ± 0.1) nm. However, the actual resonance frequencies in these microresonators scatter widely over more than 6 nm. While the first two resonators show a strong offset of 3 nm, the last five resonators each decrease their individual resonance wavelength slightly by 1 nm. On other samples, a similar behavior was observed. The measurement series clearly show that there is an inconclusive dependency between the actual size of a resonator and its spatial position on top of a microchip. The resonance frequencies widely scatter within one FSR of the system even for identical processing parameters during production. This result again validates the requirement for a wide-range tuning ability even on microresonators which are produced with highly reliable lithographic methods. It can be further assumed that at minimum a full FSR is required for tuning an arbitrary disk resonance to a fixed external frequency reference. For the investigated microresonators, a tuning scheme is required that allows for resonance shifts over a full spectral range of up to 1.3 THz.

Temperature Tuning

For optically measuring the thermal tuning abilities of disk-type silica microresonators, the devices were examined within a temperature stabilized characterization setup. It consisted of a closed loop controlled thermoelectric Peltier element and allowed to continuously set the chip temperature to arbitrary values between 15 °C and 40 °C. After changing the temperature, the samples were allowed to fully thermalize with the environment. A stabilization of the measured resonance frequency could be observed after some minutes. This allowed measuring the absolute resonance frequency at different chip temperatures. Afterwards, the measurement series was repeated for the reverse tuning direction. During the measurements, no hysteresis or other directional differences could be observed.

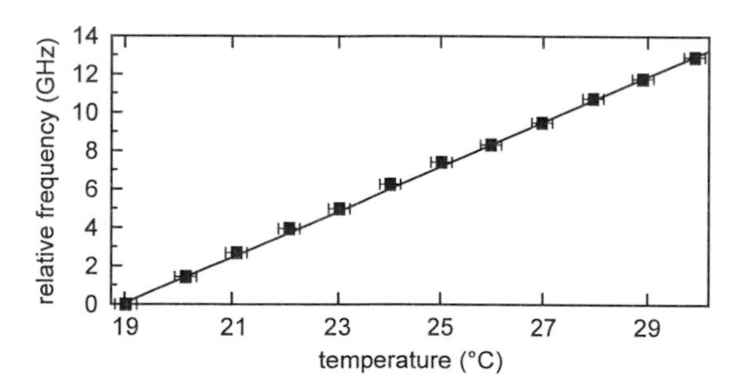

FIG. 4.10. Temperature tuning curve for a disk-type silica microresonator with a diameter of 50 µm. The temperature was varied between 19 °C and 30 °C. The system was allowed to completely settle after performing a temperature change.

With this configuration, a maximum thermal tuning of up to 40 GHz could be achieved. The corresponding calibration curve between 20 °C and 30 °C is shown in FIG. 4.10. From this measurement, an experimental tuning rate of (1.15 ± 0.04) GHz/K can be calculated. For estimating, the theoretical model introduced in Chapter 3.7.2 can be applied. By using Eq. (3.18) and the material constants of fused silica (n_{1550nm} = 1.444, $\alpha = 0.55 \cdot 10^{-6}$ K^{-1}, $\beta = 11.5 \cdot 10^{-6}$ K^{-1}) [37], [38], a theoretical shift rate of 1.65 GHz/K can be calculated. This value is slightly larger as the experimentally measured shift rate. However, by taking the higher material density of thermally grown silica into account, this result is still in good agreement with the experimental data. It is also assumed that the different expansion coefficients of the silica disk and the silicon substrate induce additional stresses inside the material.

4.4.3 Tuning Method

For applications requiring a particular WGM resonance frequency, permanent tuning by directly changing the physical dimensions of a resonator and carefully removing material from the surface is the most direct approach. In contrast to methods which add material to the resonator, etching is a simple process which allows resizing even under normal working conditions in an optical laboratory. However, such a resizing process shall not change any other of the physical resonator properties. Especially the surface roughness on the edges of the resonator is a highly crucial parameter as the Q factor of a resonant system is often limited by this feature.

Therefore, the standard dip-etching technique from silicon semiconductor processing was implemented for the resizing. This technique is well-known for the possibility to chemically polish surfaces after performing RIE processes. By shortly dipping a sample into a highly concentrated wet chemical etching solution, any sharp edges or residual surface roughnesses from ion bombardment are rounded and etched faster compared to all smoother structures on the microchip. As dip-etching with buffered hydrofluoric acid (BHF) is also used as final post-processing step during the chip production (see also Sections 4.2.4 and 4.2.5), the optical Q factors of the WGMs inside a resonator should not be influenced by this process. Therefore, this technique seems suitable for successively removing silica from the surface of a resonator even in a post-production process under laboratory conditions.

As wet chemical etching agent for silica dip-etching, BHF is widely used throughout silicon technology. This solution shows high silica etching rates and has a good selectivity against silicon. By using a buffered solution, the etching process is more controllable and constant etching rates can be achieved [39]. For the process, a BHF solution with ammonium as buffer was chosen (NH_4(conc.)/HF(conc.)/H_2O, 3:1:3). The etching rate of this BHF solution can be controlled by simply modifying the pH-value of the solution, i.e., by further diluting with water. Additional 1:10 and 1:100 etching solutions were prepared to estimate the optimum timescale for the required size change of the microresonators.

Although the final etching rates for these standard etching solutions can be found in the literature, actual results are strongly dependent on the specific chemical and physical properties of the etched silica. It can be assumed that the maximum removal rates for the different etching solutions can be observed on unstructured wafers where the silica layer is homogeneously etched. The whole etching process was performed stepwise, so the etching rates could be measured with highest precision on short or long timescales. After each etching step, the residual thickness of the silica layer was measured by white light interferometry in a wavelength range between 400 nm and 900 nm. The results of these measurements are shown in FIG. 4.11. The observed etching process is highly linear for the whole silica layer thickness. This directly proofs the high homogeneity of thermally oxidized silica. The resulting etching rates were measured dilution dependent as 20 nm/s, 2 nm/s, and 0.15 nm/s, respectively. A linear etching behavior was found for all three etching solutions.

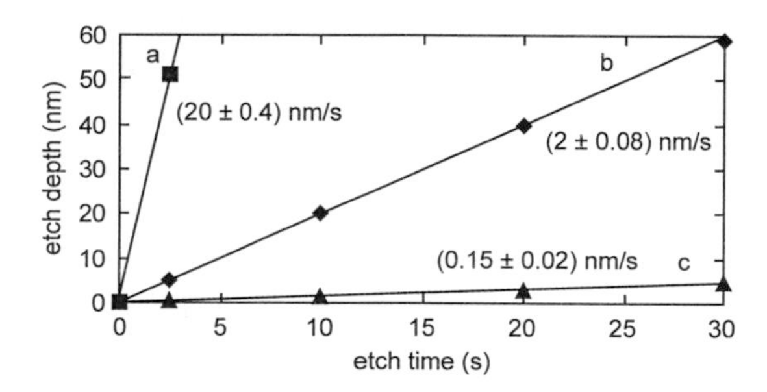

FIG. 4.11. Linear fits to the etching rates in three different BHF solutions (BHF:H_2O, a - 1:1, b - 1:10, c - 1:100). The rates were measured on unstructured silicon wafers with 2 μm thick layers of thermally oxidized silica. With this measurement, the maximum possible etching rates are assumed to be observed. The individual errors are well below the data point size.

For simple and reliable etching processes, the use of solely one solution is desired. For estimating the required etching times, maximum resonance shifts in the range of a FSR are assumed. With the simple resonance condition already presented in Chapter 3.5.1 (Eqs. (3.6) and (3.7)), it is possible to calculate the size difference which is required for shifting a WGM from one radial mode order to another. For the used disk-type silica microresonators ($r = 25$ μm, $n = 1.44402$, $\lambda = 1550$ nm), the calculated FSR is 1.3 THz. This value corresponds to a required maximum radial size reduction of 170 nm.

For the examined disk structures and a resonance wavelength of 1560 nm, a silica removal of 1 nm corresponds to a resonance shift of nearly 8 GHz. From this result, it can be concluded that the removal rate of the original BHF solution is too high for selectively removing just a few nanometers of silica. On the other hand, with the lowest dilution only shift rates below 1 GHz/s could be observed. This value is too low for a substantial resonance shift in the range of a FSR. With such a solution, extensive etching times would be required. This would highly reduce the reproducibility of the etching results and lower the achievable quality of the etching process. Furthermore, with this solution, highly inhomogeneous etching was observed on the plane surface of the wafer. This may be due to the low concentration of the active etching agents. However, solely the 1:10 etching solution resulted in a reproducible and homogenous silica removal over the whole surface of the wafer. The observed etching rate of 2 nm/s corresponds to possible resonance shift rates of 16 GHz/s. The selected dilution is a tradeoff between ease and reliability, allowing a fast single-step tuning process outside a cleanroom environment. The removal rate in the selected solution is slow enough for small resonance shifts but also suitable for larger FSR-wide tuning steps. Thus, it can be used for coarse and fine-tuning steps as well. For the selected solution, a maximum etching time of 1.5 min can be calculated for shifting a full FSR. If ultimate tunability with highest precision is required, a higher dilution of the etching solution can still be used with less reliability on the final etching result.

For determining the homogeneity of the etching process and to estimate the influence on the surface quality, the roughness of the oxide layer was analyzed before and after the etchings. A profiler (KLA-Tencor P-16) was used to measure the average surface roughness R_a of the wafers. In principal, this device is comparable to an atomic force microscope working in contact mode. A scan length of 100 µm was selected as it represents roughly the perimeter of a small microdisk. An increase of the scan length was not possible as this would also increase the chance to catch up background noise from the cleanroom. The scan speed was set to 2 µm/s with a nominal resolution of 0.0078 Å/point. The maximum vertical scan range was 13 µm. With these parameters, an average surface roughness $R_a \approx 1.5$ nm was measured on unprocessed wafers. After dip-etching 600 nm of silica, a reduced surface roughness $R_a' \approx 1.4$ nm was achieved. This result verifies that etching with the selected BHF solution does not significantly influence the observable surface roughness. The measured average roughness of the etched surface corresponds well with other experimental data also determined for the surface quality of directly etched microresonators [26], [36]. This directly proves the comparability of the resulting Q factors in microdisks.

By using a wet chemical etching solution, it can be assumed that the silica will be dissolved highly isotropic from all sides. Thus, the etching process will not only include the outer edges of the resonators but also their upper and lower sides. For full FSR-wide resonance tuning, the original silica layer thickness of 2 µm would be reduced to nearly 1.7 µm. For illustrating the influence of this effect, in FIG. 4.12a the energy distribution of the fundamental resonator mode is shown. It was calculated with a numerical mode solver (Photon Design Fimmwave).

Although the difference between the two silica thicknesses is nearly 10 % before and after etching, the modal properties of the observed WGMs are basically unchanged. By thinning the silica layer, the field maximum of the mode is slightly shifted towards the center of the disks. This can be seen in FIG. 4.12b where a horizontal cut through the middle of the energy distributions of the fundamental modes in microdisk resonators with silica thicknesses of 1.7 µm and 2 µm are presented. The maximum of the field distribution is slightly shifted to the center of the disk by only 5 nm when the silica thickness is reduced in correspondence to a full FSR-wide tuning step. By assuming that this shift directly relates to the effective refractive radius r_{eff} of the fundamental mode, this will additionally add to the intended size reduction from etching. In the case of full FSR-wide tuning, this effect will cause a maximum deviation of 3 % in the relation between the observed resonance shift and the corresponding size change. Hence, at the given wavelengths around 1560 nm, the effect of silica layer thinning can be neglected for estimating the required etching times.

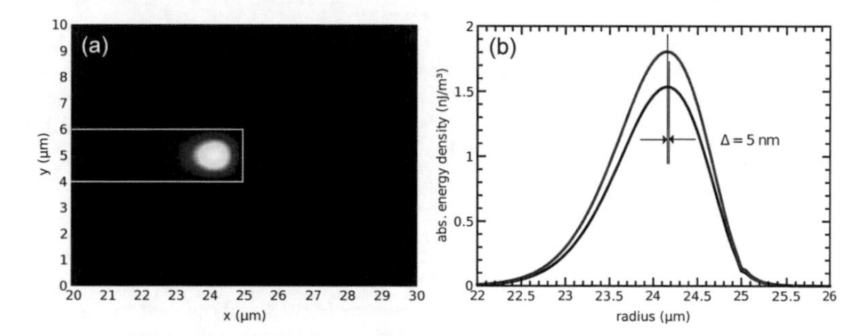

FIG. 4.12. (a) Absolute energy distribution of the fundamental mode in a disk-type silica microresonator with a radius of 25 µm. The thickness of the silica layer is 2 µm. (b) The difference of the effective radius r_{eff} between the mode shown in (a) and a similar mode in a resonator with same dimensions but reduced silica layer thickness of 1.7 µm is plotted. The effective radius of the mode is nearly unaffected by this structural change. The given energy densities are arbitrary and uncorrelated.

4.4.4 Results

After exactly measuring the initial optical properties of the resonators, the complete microchip was fully emerged into the selected etching solution for some seconds. Afterwards, the microchip was rinsed by purified water to stop any residual etching processes and a new measurement of the resonances was performed. By repeating this process, the statistics of the etching approach could be measured. For measuring the reliability of the process, two resonators on the same microchip were simultaneously analyzed. There initial frequency difference was used to select two specific HCN reference lines with comparable difference. With the P16 and P18 reference lines, the frequency offset between the two modes and their corresponding references was around 1 THz and thus well below the FSR of the resonant system. Resonance tuning was then performed by applying multiple repeated etching steps.

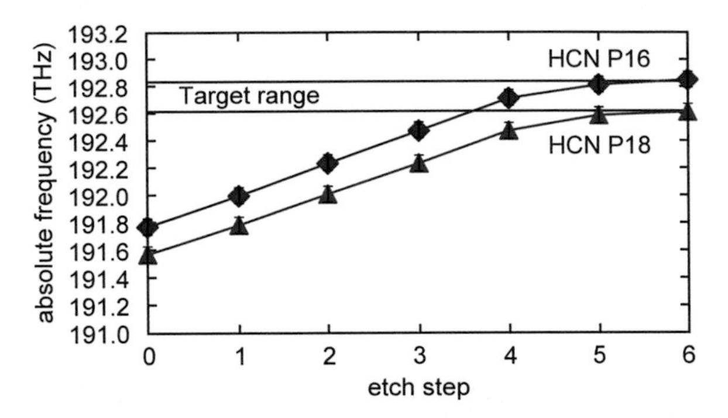

FIG. 4.13. Simultaneous post-production wet chemical etch tuning of two different resonators located on the same microchip. The graphs show the reproducibility of the different etching steps and between consecutive steps with identical processing parameters. The first four steps had each a duration of 60 s, the following two etching steps had reduced etching durations of 30 s and 6 s.

In FIG. 4.13, the results of such a measurement series are shown. In summary, six independent etching steps were performed. The first four had each a duration of 60 s, the following two steps had reduced etching durations of 30 s and 6 s. During all steps, a constant frequency offset of 200 GHz was observed between the resonances in both resonators. This shows that highly parallel processing could be realized. With this sequential method, it was possible to achieve frequency matching between the two resonances and the pre-selected P18 and P16 HCN reference lines with a residual frequency mismatch of 7.76 GHz and 14.66 GHz, respectively.

Tuning Rate

For the used disk-type microresonators, a highly linear tuning behavior with a linear shift rate of (3.77 ± 0.03) GHz/s was observed. This tuning rate was nearly independent of the applied etching time or the number of applied etching steps. Thus, several short steps can be combined with high reliability to a single extended etching step. The applied process for permanent post-production tuning was analyzed for etching times between a few seconds and up to two minutes. In the case of very short etching times (< 10 s), slightly increased tuning rates could be observed. This effect can be explained by the consumption of the initial etching agents inside the solution. In a steady-state etching process, the reaction products have to be constantly removed from the surface and replaced by new reactants. This can be done by stirring or constantly moving the sample. On short timescales, higher etching rates are observed. This is because the initial concentration of reactants is higher as in the equilibrium state and no previously produced reaction products have to be replaced.

Based on theoretical considerations, the observed resonance shift rate corresponds to a mean disc radius removal rate of 0.5 nm/s. This result deviates from the observed etching rate on unstructured wafers by a factor of 4. The reason for this difference is unknown. The outer silica layer may still be passivated from the plasma etching process. As passivation is a surface effect which can substantially lower the observed etching rate, an extended etching over more than a few hundred nanometers of silica should show a strong influence on the observed etching rates. However, this could not be verified with the actual design because of the increasing influence of a reduced silica layer thickness to the modal structure. After extended etching processes, the resonance frequency of a mode can not be used anymore for measuring the actual size of a resonator. Another explanation for the observed etching time difference between unstructured and structured wafers could be a possible influence of the height to length ratio by etching curved structures. Also the high curvature of the edges in disk-type microresonators could affect the observed etching rates. However, a detailed analysis of this effect was not performed as even for an FSR-wide tuning no relevant changes in the etching rate could be observed.

Tuning Precision

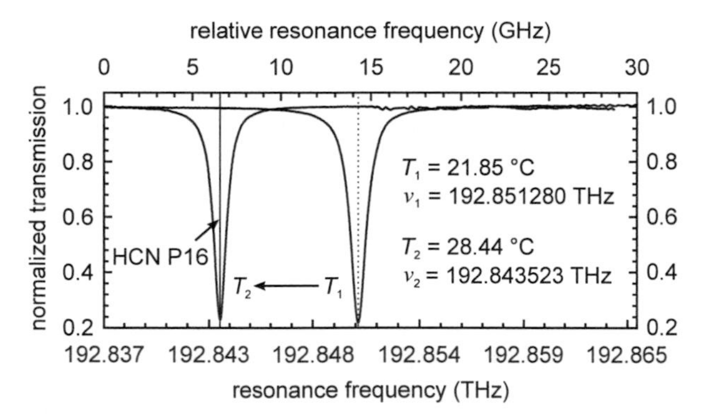

FIG. 4.14. After approaching a specific reference line (HCN P16), final tuning within a few GHz was performed by thermally shifting the resonance with the Peltier element from the temperature control setup.

With the described technique, highly reproducible tuning could be demonstrated. The WGM resonances could be tuned within 10 GHz to specific pre-defined frequencies, i.e., two HCN reference lines (P18 and P16). Thereby, even with extended etching times, a good reproducibility of the etching results could be observed. The high reliability of the process even allows for tuning with comparable precision over a full FSR in a single etching step. The observed minimum tuning is in the range of 10 GHz or even below. It is limited by the possibility of an exact control over the active etching times. Even for short dip-etches, the dissolving reaction sustains for at least 2-3 s in the selected solution. Thus, minimum layers of around 1 nm are always removed from the structure. If ultimate etching accuracy is required, a second etching solution with a higher dilution can be used. It reduces the effective etching rates and less material is removed per time. However, this also lowers the reliability of the process and reduces the predictability of the final etching rate. For ultimate fine-tuning, this effect may still be negligible over short frequency ranges.

The observed minimum tuning range does not allow a perfect matching between WGM resonance and external frequency references. As the measured resonance frequency of a resonator also depends on the physical parameters of the environment, such perfect permanent tuning is not required for practical applications and an active method for small residual fine-tuning is more essential.

In this experiment, temperature tuning (see Section 4.4.2) was applied. Microresonator applications often required some sort of temperature stabilization. This control stage can also be used for regulating the temperature of the system. As a proof-of-principle, such a control scheme was implemented for tuning the residual mismatch of the observed WGM resonances into perfect overlap to their corresponding reference lines. In FIG. 4.14, the Lorentzian shaped dip of the WGM with the lower residual mismatch to the corresponding HCN reference line (P16, $\Delta\upsilon$ = 7.76 GHz) is shown before and after temperature tuning. By heating the microchip from room temperature at 21.85 °C to an increased temperature of 28.44 °C, this external HCN reference could be perfectly matched. Thus, temperature tuning is an appropriate method for achieving ultimate tunability after permanent tuning by dip-etching was applied.

Quality Factor

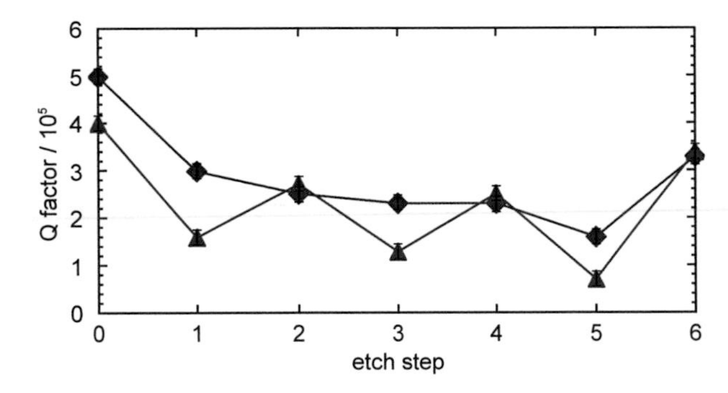

FIG. 4.15. Measured Q factors for the two observed microresonators after the six consecutive etching steps described in the caption of FIG. 4.13. The straight lines are guides to the eye.

The six repeated etching steps did not change the measured Q factors of the observed resonances. Throughout the complete measurement series, the Q values remained in the mid 10^5 range as shown in FIG. 4.15. For both microresonators, an equivalent linewidth behavior was observed. The measurements proof the previous results on unstructured wafers and show that the surface quality of the resonator edges is unaffected by additional dip-etching processes. Thus, the surface quality of a resonator remains on the same level as it was achieved directly after production. It can be assumed that the high quality of the resulting surface finish after etching can also be observed on disk-type or wedge microresonators with different sizes.

4.4.5 Summary and Outlook

The presented etching tuning method offers an efficient post-production technique for permanently shifting the WGM resonances of on-chip disk-type silica microresonators. Thereby, the optical properties of the resonators are preserved. The observed Q factors of the WGM resonances are just slightly affected by the tuning process. The presented method complements standard production schemes and is also compatible with a mass production of quasi-planar photonic structures.

By dipping the microchips into a diluted BHF solution, the size of the resonators could be reduced. For tuning the microresonators, a single etching solution was selected. It allowed a high reproducibility of the process on short and long time scales. After determining the etching rates on unstructured and structured wafers, with this solution a highly linear resonance shift rate of (3.77 ± 0.03) GHz/s was measured. By considering the correspondence between the WGM resonance frequency and the geometry of the resonators, this tuning rate directly corresponds to a material removal rate of 0.5 nm/s. The high process reliability allows a controllable minimum tuning of around 10 GHz. It is also possible to tune the individual resonances with comparable precision over a full FSR within single etching steps.

An ultra-precise fine-tuning for cancelling out any residual frequency mismatch can finally be done by using other tuning techniques, e.g., temperature fine-tuning. With such a combined technique, the analyzed resonances could be perfectly matched to external HCN frequency reference lines. In contrast to permanent tuning, this additional fine-tuning also allows to take external environmental influences into account. Furthermore, this method allows the dynamical scanning of a WGM resonance over such an external frequency reference. Also individual temperature tuning of different resonators can be achieved by adding local heating elements to the chip-based design. With such an approach, tunable resonator-based photonic molecules can be realized by directly coupling multiple microdisks with pre-defined resonances.

The presented tuning method should be applicable for nearly all types of on-chip silica microresonators. All directly produced structures, like wedge and disk-type, can be processed. Toroidal microresonators may also be tuned, but a high degradation of the observed Q factors must be expected due to the reduced surface quality. The presented method may also be applied to other material systems containing comparable WGM resonator designs. In this case, a modified etching solution has to be used. All described techniques can be correspondingly applied.

4.5 Quantum Optics on Chip

Within the last section, the production and tuning of chip-based disk-type high-Q silica microresonators could be successfully demonstrated. The range of possible applications is extremely wide for these cavities; it includes classical photonic applications in telecommunication and signal processing [40], allows applications as integrated sensor and detection devices [41], and reaches also to possible future applications within the fields of quantum information processing [42] and as building blocks for quantum networks [43]. However, up to now, these non-classical applications are still matter of research. For studying cavity quantum electrodynamics (CQED), the high Q factors and small mode volumes of these cavities are essential properties. By relying on chip-based systems, in contrast to modular designs, the achievable level of integration can be highly increased. This also helps to isolate quantum systems from the environment and allows better production reliability. In CQED, the experiments are often performed under cryogenic conditions. This gives additional constructional constraints to the implemented designs. Furthermore, by working with chip-based devices, the scalability to more complex and advanced microstructures is already implemented.

In the following section, the experimental realization of on-demand positioning of preselected quantum emitters to fiber-coupled toroidal microresonator systems is briefly introduced. It proves the applicability of chip-integrated microresonators for CQED applications and is a basic proof-of-principle for further research within this field. The fine-tuning methods presented in the last sections were not applied, but they will be fundamental when the emission properties of the quantum emitters have to be matched with specific cavity resonances. The presented works were achieved in close cooperation with the Paul-Drude-Institut, Leibniz-Institut für Festkörperelektronik (PDI) and resulted in a publication [2].

4.5.1 Resonant Structures

In CQED, the interaction between specific quantum emitters and light fields, which are confined in a reflecting cavity, are investigated. If the emitter shows a simple 2-level electronic structure, the system can be described by the Jaynes-Cummings model [44]. The interaction is characterized by so-called vacuum Rabi oscillations. They describe a damped energy oscillation between the emitter and single modes of the cavity field.

Spontaneous emission of the emitter results from coupling to vacuum fluctuations of the cavity field. If the cavity is on resonance with this emission, coherent interactions between qubit states can be observed. This allows interfacing the "flying" photon states with more "stationary" quantum emitter states. It is also possible to create entanglement between the emitter and the cavity field.

In the presented experiment, the resonant cavity was a toroidal microresonator [45]. As 2-level quantum emitter systems, nitrogen-vacancy (N-V) defect centers in diamond nanocrystals (DNCs) were applied. These single-photon emitters are photo-stable even under room temperature conditions. However, using them under cryogenic conditions is highly preferred due to reduced phonon interactions and a better visibility of their zero-phonon line (ZPL). N-V defect centers in DNC are ideally suited for applications in nanophotonics and quantum technologies [46]. In contrast to single molecules, atoms, or colloidal quantum dots, diverse solid-state nanomanipulation techniques can be applied [47].

Toroidal Microresonator

For the experiments, toroidal microresonators with a diameter of 65 μm were produced with the methods described in Section 4.2. Fiber tapers with diameters around 1 μm were used for evanescently coupling to the WGM resonances. For resolving the high-Q resonances of the cavity, a tunable Ti:sapphire laser running at a central wavelength of 788.75 nm and with a maximum scanning range of 20 GHz was applied. With this configuration, Q factors in the mid 10^5 range could be observed. These results were mainly limited by imperfections induced during production; on later samples Q factors within the upper 10^6 range could be measured. Within the accessible wavelength range of the laser, the resonators have a calculated FSR of 2.1 nm. This allowed observing a multitude of different mode orders within the scanning range of the laser.

Nitrogen-Vacancy Defect Centers in Diamond

The diamond crystal lattice can be distracted by impurities replacing original carbon atoms. This naturally occurring effect is the reason for different colors in diamond. Many defects are luminescent and posses discrete electronic energy states [48]. Most common are the N-V defect centers where two carbon atoms are replaced by a nitrogen atom and a free lattice site [49].

The structure of such N-V defects is shown in FIG. 4.16a. They can be observed as neutral ($N\text{-}V^0$) and negatively charged ($N\text{-}V^-$) defects. For the $N\text{-}V^-$ center, additional electrons are involved in the electronic configuration. These electrons can be provided from other nitrogen atoms nearby the defect. Thus, the absolute nitrogen density affects which type of defect center is predominant within the crystal [50], [51]. The $N\text{-}V^-$ shows a ZPL around 637 nm from a transition between two energetic triplet states [52].

This can be seen from FIG. 4.16b where the room temperature emission spectrum of a single N-V⁻ defect center is shown. Often both N-V center configurations are present which indicates a possible conversion between the $N\text{-}V^0$ and $N\text{-}V^-$ states. The defect centers are quasi atomic few-level systems on which non-classical single-photon emission can be observed [48]. The defect centers can be non-resonantly excited by lasers emitting at wavelengths around 532 nm.

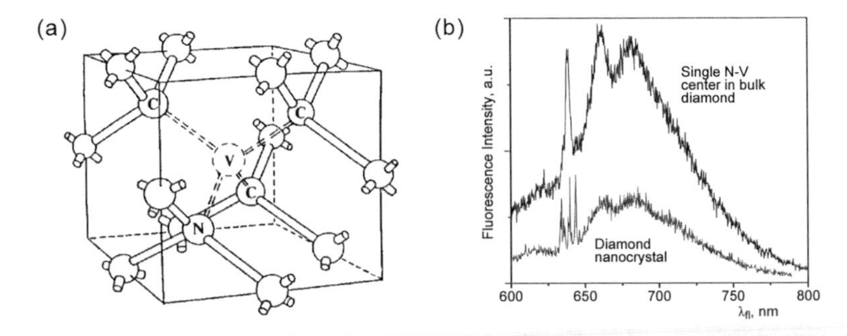

FIG. 4.16. N-V centers in diamond (from [53]). (a) Elementary cell of the diamond crystal lattice containing a single N-V defect center. Two carbon atoms are replaced by a nitrogen atom and a free lattice site. (b) Room temperature emission spectrum of single N-V centers in bulk diamond and multiple N-V centers in nanocrystalline diamond. Narrow ZPLs with alternated phonon sidebands are observed.

For the presented experiment, a nitrogen-rich industrial diamond powder solution containing DNCs with diameters around 100 nm (MicroDiamant, Switzerland) was used. In the DNCs, the $N\text{-}V^-$ defects are predominant and strong ZPLs around 637 nm can be observed in the corresponding spectra. As only 10 % of the nanodiamonds contain N-V defect centers, a manual preselection process by measuring the individual photoluminescence has to be performed in advance. However, due to the large size of the DNCs, multiple N-V centers can be located inside a DNC. For discriminating between multiple and single photon emitters, the photon statistics of a luminescent DNC can be determined by an intensity autocorrelation measurement in a so-called Hanbury Brown and Twiss (HBT) setup (for details about photon statistics and measurement see Appendix C) [54]. With this technique, it is also possible to estimate the number of defect centers which are contained within a single DNC.

4.5.2 Experimental Methods

The aim of the presented experiment was to demonstrate the possibility of on-demand positioning of preselected DNCs containing N-V defect centers onto the outer rim of high-Q toroidal microresonators. Thereby, the efficient coupling between these two optical constituents had to be proven. As DNCs are fully crystalline structures with irregular rough surfaces, applicability to the smooth surface of the resonators had to be verified and evanescent coupling between defect centers and resonator modes should be demonstrated.

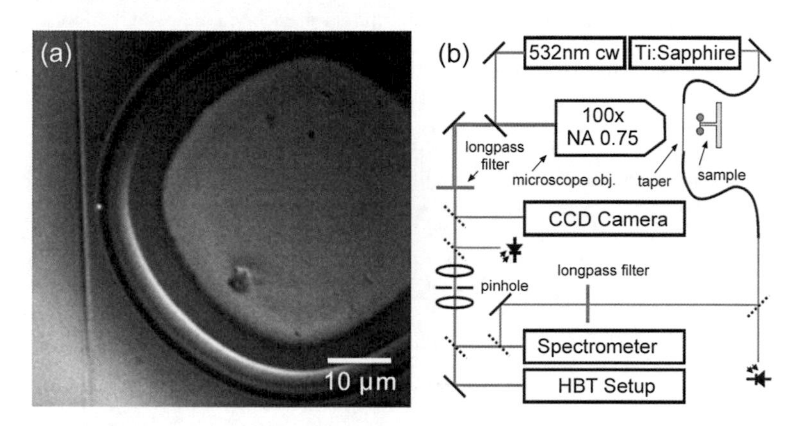

FIG. 4.17. (a) Microtoroid with a 65 µm diameter having a preselected DNC on the outer rim. The DNC is optically excited and can be seen by its scattering. The vertical line in the left corner is the used fiber taper. (b) Experimental setup for characterizing the coupled resonator system. It contains a homebuilt optical microscope with confocal operation mode, excitation lasers, a spectrometer and a Hanbury Brown and Twiss (HBT) setup.

For the required preselection process of the DNCs, a stripped optical fiber was used as material reservoir. Multiple single DNCs could be deposited on the outside of this fiber by dip-coating into the diluted diamond solution. Moving and rotating the fiber within the focus of a wide-field optical microscope allowed a one-by-one analysis of the DNCs. By using the confocal mode of the optical microscope, the optical properties of different DNCs were measured and a suitable emitter was selected. Afterwards, this emitter could be picked up by a tapered transfer fiber with only a few micrometers in diameter. It was placed perpendicular to the selected DNC in the focus of the confocal microscope such that the complete transfer process could be monitored. By replacing the donor fiber with a microtoroid, the same procedure could be done in reverse. With this technique, the selected DNC can be observed in the microscope without loosing the actual position of the crystal. The selected DNC was then placed within the maximum of the evanescent field of the fundamental resonator mode, i.e., directly at the outer rim of the toroid in the equatorial plane of the resonator. Optical identification of this region was performed by interpreting the depth of focus of the used microscope objective. A further description of this technique can be found in [2] and [55].

After removing the transfer fiber, a single DNC could be successfully attached to the toroidal microresonator. The DNC can be seen in FIG. 4.17a. For measuring the optical properties of the combined system, an experimental setup as depicted in FIG. 4.17b was used. It allows observing the WGMs of the resonator by using an evanescently coupled fiber taper or by collecting the light which is scattered into the direction of the microscope objective. Furthermore, it is possible to use the microscope to confocally excite the applied DNC and to collect its photoluminescence via the resonator-coupled fiber taper. For this, the optimum coupling position of the fiber taper had to be found in advance. This was done by first exciting a WGM with the fiber taper and then changing its coupling position to maximize the intensity of light which is scattered out of the resonator. After the optimum position was found, excitation and collection paths were changed such that the fiber taper could be used to couple light out of the resonator. For detection, a spectrometer and a HBT setup were integrated within the setup to analyze the collected light.

In order to demonstrate the feasibility of the coupled CQED system also for cryogenic applications, the resonator system was then mounted inside a cold-finger cryostat and cooled to liquid helium temperature. In a cryostat, the optical emission of N-V defect centers can be observed via enhanced ZPLs without broadening. The used cryostat setup did not allow the integration of fiber tapers or other coupling devices. Therefore, only standard luminescence measurements were performed in the cryostat.

4.5.3 Results

The result of spectroscopic measurements on the attached DNC is shown in FIG. 4.18a. The red curve is the unmodulated optical spectrum of the attached DNC by confocal excitation and collecting the fluorescence via the microscope. There is no noticeable influence of the resonator observable and the obtained spectrum corresponds to reference spectra measured on free diamond samples. For the blue curve, the DNC was also confocally excited, but the fluorescence was collected via the fiber taper. The strong modulation in the observed spectrum reveals coupling of the diamond emission to the high-Q modes of the microresonator. A further analysis of the observed spectrum and the contained periods of modulation shows good agreement with the FSR of the fundamental WGM in the used microresonator [2], [55]. The intensities of the two curves are scaled to the same level; the measured ratio was 1:15 for the taper-collected signal. Therefore, the lack of modulation in the directly measured spectrum shows that light which was already coupled into the WGM is just moderately scattered by the attached DNC [56]. However, the observed intensity relation between both curves show that most of the photoluminescence is emitted into free space and only moderate coupling to the resonator modes could be achieved. By taking the relatively large size of the used DNCs into account, the coupling can be further optimized by increasing the overlap between the evanescent field of the WGM and the absolute position of the DNC. By using smaller nanocrystals and microresonators with smaller radii, their mutual interaction is increased and higher coupling efficiencies can be achieved.

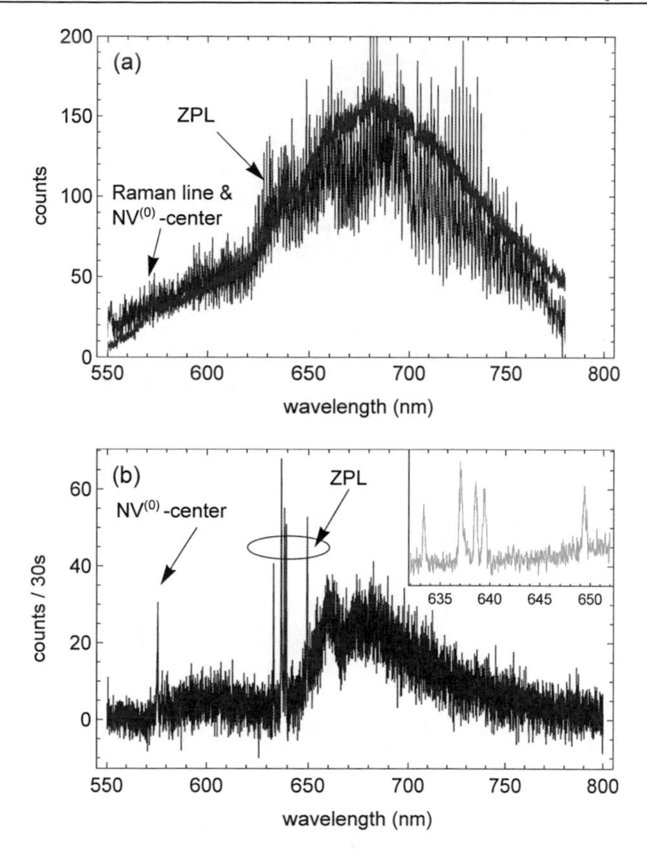

FIG. 4.18. (a) Room temperature photoluminescence spectra of the attached DNC. The red curve corresponds to direct collection of free space radiation whereas the blue curve is collected via the resonator coupled fiber taper. The modulation corresponds to the high-Q modes of the resonator. For the same integration time of the measurement, the blue curve is scaled by a factor of 15. (b) The same spectrum taken at liquid helium temperatures. The narrowing of the original ZPL peak reveals the existence of five individual ZPLs. The inset shows a zoom-in around the ZPLs.

The microtoroid and the attached DNC were both cooled to liquid helium temperature. This allows obtaining the cryogenic emission properties of the N-V defect centers by simply measuring the photoluminescence spectrum of the DNC. The result is presented in FIG. 4.18b. By cooling, the phonon sidebands of the N-V defect centers are highly suppressed and individual ZPLs become visible due to a narrowing in the spectral linewidth of the defect center transition. For the attached DNC, five individual ZPLs are distinguishable from their spectral peaks. The asymmetric line shape at 637.5 nm may also indicate a spectral line doublet which can not be resolved by the used spectrometer. Thus, for the selected DNC, at least five N-V centers with different spectral positions were found. For controlled CQED experiments, single N-V defect centers have to be accessed. Therefore, DNCs containing single N-V centers are preferred. If several individual defect centers are contained within a single DNC, these centers may emit at slightly different ZPL frequencies. If the gap between these lines is spectrally resolvable, then accessing single N-V centers may also be possible in DNCs containing several of such defects.

For further analyzing the statistics of photons emitted by the individual N-V defect centers, the second order intensity correlation function $g^{(2)}(\tau)$ was measured with the integrated HBT setup (see also Appendix C). The result of this measurement is presented in FIG. 4.19. The obtained data shows a clear antibunching behavior for zero time difference τ. This reveals the non-classical light emission of the N-V centers contained inside the DNC. From the depth of the antibunching dip, the number of contained single photon emitters can be estimated [57], [58]. The $g^{(2)}$-function of N individual photon emitters is given by $g^{(2)}(\tau) = 1 - 1/N \exp(-k|\tau|)$ with $k = k_{exc} + k_{em}$ for the excitation and emission rates of the individual N-V defect centers in the DNC. By fitting this function to the acquired statistical data, $N = 6.2 \pm 0.3$ is obtained for the number of contained defects. This indicates that at least six N-V defect centers were located inside the selected DNC [57], [58]. The result is in good agreement with the spectral data measured at cryogenic temperatures.

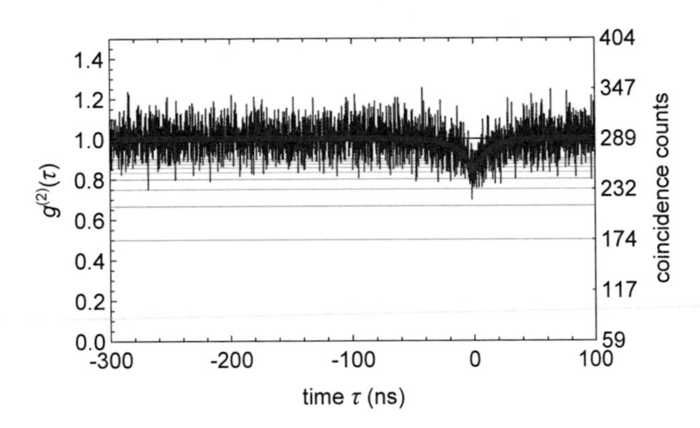

FIG. 4.19. Second order intensity correlation function $g^{(2)}(\tau)$ of the spectral emission of the DNC attached to the microtoroid. A clear antibunching behavior is observed at zero time difference. From the depth of the dip, the number of individual single photon emitters N can be estimated. The horizontal lines are spaced in $1-1/N$ distance. This indicates that at least six N-V defect centers are contained within the selected DNC.

4.5.4 Summary and Outlook

The results clearly show that the fluorescence of N-V defect centers in DNCs can be efficiently coupled to the high-Q resonances of chip-based toroidal microresonators. The presented pick-and-place technique allows an efficient preselection of suitable DNCs with just a single or a low number of N-V centers inside the crystal. Due to the possibility of continuously monitoring the process under a microscope, the nanometer sized structures can be handled and placed on-demand. By measuring the emission properties of the DNC in free space and after transmission through the resonator, the coupling of the phonon sideband to individual WGM modes could be demonstrated. The measurement of the photon statistics and the emission spectrum at cryogenic temperatures revealed the existence of at least six individual defect centers within the selected DNC. The ZPLs of these defect centers showed slightly different emission frequencies and could be spectrally resolved. This is analog to experiments already done with single molecules inside a cryostat [59]. The occurrence of frequency spacing between different ZPLs within a single DNC relaxes the requirements for the realization of controlled CQED experiments. Even with DNCs containing multiple N-V centers, in a cryostat single defects can be selected out of an ensemble by their specific ZPL frequencies. Furthermore, individual N-V defect centers can be addressed and matched with specific WGM resonances by tuning the resonators.

With these investigations, the general applicability of chip-based microresonators for CQED experiments could be successfully demonstrated. By providing the manipulation techniques for placing individual preselected DNCs on demand, first CQED designs can already be realized with this approach. For future experiments, the mutual coupling between N-V defect centers and resonator modes has to be increased by using smaller DNCs and resonator structures. For a controlled CQED experiment, also the transition frequency of a single ZPL needs to be matched with the resonances of the WGMs. For coarsely pairing both frequencies, a permanent tuning method can be applied. The selective etching approach presented in the previous section may be feasible. For an additional active tuning of the two resonances, other non-permanent tuning methods must be applied. A suitable selection can be found in Section 4.3.

[1] R. Henze, C. Pyrlik, A. Thies, J. M. Ward, A. Wicht, and O. Benson, "Fine-tuning of whispering gallery modes in on-chip silica microdisk resonators within a full spectral range," *Appl. Phys. Lett.*, vol. 102, no. 4, p. 041104, 2013.

[2] M. Gregor, R. Henze, T. Schröder, and O. Benson, "On-demand positioning of a preselected quantum emitter on a fiber-coupled toroidal microresonator," *Appl. Phys. Lett.*, vol. 95, no. 15, p. 153110, 2009.

[3] T. Schröder, "Herstellung und Vermessung von toroidalen Mikroresonatoren," Humboldt-Universität zu Berlin, 2007.

[4] S. J. Moss and A. Ledwith, Eds., *Chemistry of the Semiconductor Industry*. New York: Chapman and Hall, 1987.

[5] S. Wolf and R. N. Tauber, *Silicon Processing for the VLSI Era: Volume 1 - Process Technology*, 2nd ed. Sunset Beach: Lattice Press, 2000.

[6] H. Janseny, H. Gardeniers, M. de Boer, M. Elwenspoek, and J. Fluitman, "A survey on the reactive ion etching of silicon in microtechnology," *J. Micromechanics Microengineering*, vol. 6, no. 1, pp. 14–28, 1996.

[7] S. Ray, C. Maiti, and S. Lahiri, "Chemically assisted ion beam etching of silicon and silicon dioxide using SF_6," *Plasma Chem. Plasma Process.*, vol. 15, no. 4, pp. 711–720, 1995.

[8] J. Pelletier, Y. Arnal, and A. Durandet, "SF_6 Plasma Etching of Silicon: Evidence of Sequential Multilayer Fluorine Adsorption," *Europhys. Lett.*, vol. 4, no. 9, pp. 1049–1054, 1987.

[9] D. K. Armani, T. J. Kippenberg, S. M. Spillane, K. J. Vahala, C. Briggs, T. Buxkemper, L. Czaia, H. Green, and S. Gustafson, "Ultra-high-Q toroid microcavity on a chip," *Nature*, vol. 421, no. 6926, pp. 925–928, 2003.

[10] M. Cai, O. Painter, and K. J. Vahala, "Observation of critical coupling in a fiber taper to a silica-microsphere whispering-gallery mode system," *Phys. Rev. Lett.*, vol. 85, no. 1, pp. 74–77, 2000.

[11] T. J. Kippenberg, S. M. Spillane, D. K. Armani, B. Min, L. Yang, and K. J. Vahala, "Fabrication, coupling and nonlinear optics of ultra-high-Q Micro-Sphere and chip-based toroid microcavities," in *Optical Microcavities*, K. J. Vahala, Ed. Singapore: World Scientic, 2004, pp. 177–238.

[12] J. Rezac and A. Rosenberger, "Locking a microsphere whispering-gallery mode to a laser," *Opt. Express*, vol. 8, no. 11, pp. 605–610, 2001.

[13] P. E. Barclay, C. Santori, K.-M. Fu, R. G. Beausoleil, and O. Painter, "Coherent interference effects in a nano-assembled diamond NV center cavity-QED system," *Opt. Express*, vol. 17, no. 10, pp. 8081–8097, 2009.

[14] Y.-S. Park, A. K. Cook, and H. Wang, "Cavity QED with diamond nanocrystals and silica microspheres," *Nano Lett.*, vol. 6, no. 9, pp. 2075–2079, 2006.

[15] W. von Klitzing, R. Long, V. S. Ilchenko, J. Hare, and V. Lefèvre-Seguin, "Frequency tuning of the whispering-gallery modes of silica microspheres for cavity quantum electrodynamics and spectroscopy," *Opt. Lett.*, vol. 26, no. 3, pp. 166–168, 2001.

[16] C. Ciminelli and C. Campanella, "Label-free optical resonant sensors for biochemical applications," *Prog. Quantum Electron.*, vol. 37, no. 2, pp. 51–107, 2013.

[17] T. Ibrahim, W. Cao, Y. Kim, J. Li, J. Goldhar, P.-T. Ho, and C. Lee, "All-optical switching in a laterally coupled microring resonator by carrier injection," *Photonics Technol. Lett.*, vol. 15, no. 1, pp. 36–38, 2003.

[18] T. Sadagopan, S. J. Choi, S. J. Choi, K. Djordjev, and P. D. Dapkus, "Carrier-induced refractive index changes in InP-based circular microresonators for low-voltage high-speed modulation," *Photonics Technol. Lett.*, vol. 17, no. 2, pp. 414–416, 2005.

[19] A. Guarino, G. Poberaj, D. Rezzonico, R. Degl'Innocenti, and P. Günter, "Electro-optically tunable microring resonators in lithium niobate," *Nat. Photonics*, vol. 1, pp. 407–410, 2007.

[20] T. Ioppolo, U. Ayaz, and M. V. Otügen, "Tuning of whispering gallery modes of spherical resonators using an external electric field," *Opt. Express*, vol. 17, no. 19, pp. 16465–16479, 2009.

[21] M. Humar, M. Ravnik, S. Pajk, and I. Muševič, "Electrically tunable liquid crystal optical microresonators," *Nat. Photonics*, vol. 3, no. 10, pp. 595–600, 2009.

[22] D. Armani, B. Min, A. Martin, and K. J. Vahala, "Electrical thermo-optic tuning of ultrahigh-Q microtoroid resonators," *Appl. Phys. Lett.*, vol. 85, no. 22, p. 5439, 2004.

[23] J. M. Ward and S. N. Chormaic, "Thermo-optical tuning of whispering gallery modes in Er:Yb co-doped phosphate glass microspheres," *Appl. Phys. B*, vol. 100, no. 4, pp. 847–850, 2010.

[24] A. Chiba and H. Fujiwara, "Resonant frequency control of a microspherical cavity by temperature adjustment," *Jpn. J. Appl. Phys.*, vol. 43, no. 9A, pp. 6138–6141, 2004.

[25] Q. Ma, T. Rossmann, and Z. Guo, "Whispering-gallery mode silica microsensors for cryogenic to room temperature measurement," *Meas. Sci. Technol.*, vol. 21, no. 2, p. 025310, 2010.

[26] H. Lee, T. Chen, J. Li, K. Y. Yang, S. Jeon, O. Painter, and K. J. Vahala, "Chemically etched ultrahigh-Q wedge-resonator on a silicon chip," *Nat. Photonics*, vol. 6, no. 6, pp. 369–373, 2012.

[27] K. Srinivasan and O. Painter, "Optical fiber taper coupling and high-resolution wavelength tuning of microdisk resonators at cryogenic temperatures," *Appl. Phys. Lett.*, vol. 90, no. 3, p. 031114, 2006.

[28] V. S. Ilchenko, P. S. Volikov, V. L. Velichansky, F. Treussart, V. Lefèvre-Seguin, J.-M. Raimond, and S. Haroche, "Strain-tunable high-Q optical microsphere resonator," *Opt. Commun.*, vol. 145, no. 1, pp. 86–90, 1998.

[29] Y. Louyer, D. Meschede, and A. Rauschenbeutel, "Tunable whispering-gallery-mode resonators for cavity quantum electrodynamics," *Phys. Rev. A*, vol. 72, no. 3, p. 031801, 2005.

[30] S. I. Shopova, G. Farca, A. T. Rosenberger, W. M. S. Kotov, A. Wickramanayake, and N. A. Kotov, "Microsphere whispering-gallery-mode laser using HgTe quantum dots," *Appl. Phys. Lett.*, vol. 85, no. 25, pp. 6101–6103, 2004.

[31] H.-J. Moon, Y.-T. Chough, and K. An, "Cylindrical Microcavity Laser Based on the Evanescent-Wave-Coupled Gain," *Phys. Rev. Lett.*, vol. 85, no. 15, pp. 3161–3164, 2000.

[32] P. Tamarat, T. Gaebel, J. R. Rabeau, M. Khan, A. D. Greentree, H. Wilson, L. C. L. Hollenberg, S. Prawer, P. Hemmer, F. Jelezko, and J. Wrachtrup, "Stark shift control of a single optical center in diamond," *Phys. Rev. Lett.*, vol. 97, no. 8, p. 083002, 2006.

[33] L. C. Bassett, F. J. Heremans, C. G. Yale, B. B. Buckley, and D. D. Awschalom, "Electrical Tuning of Single Nitrogen Vacancy Center Optical Transitions Enhanced by Photoinduced Fields," *Phys. Rev. Lett.*, vol. 107, no. 26, p. 266403, 2011.

[34] B. Peng, S. K. Ozdemir, J. Zhu, and L. Yang, "Photonic molecules formed by coupled hybrid resonators," *Opt. Lett.*, vol. 37, no. 16, pp. 3435–3437, 2012.

[35] J. F. Donegan, M. Gerlach, and Y. P. Rakovich, "Symmetric photonic molecules formed from coupled microspheres," in *Proceedings of SPIE*, 2008, vol. 6872, p. 68720I.

[36] I. M. White, N. M. Hanumegowda, H. Oveys, and X. Fan, "Tuning whispering gallery modes in optical microspheres with chemical etching," *Opt. Express*, vol. 13, no. 26, pp. 10754–10759, 2005.

[37] I. Malitson, "Interspecimen comparison of the refractive index of fused silica," *J. Opt. Soc. Am. B*, vol. 55, no. 10, pp. 1205–1208, 1965.

[38] J. Matsuoka, N. Kitamura, S. Fujinaga, T. Kitaoka, and H. Yamashita, "Temperature dependence of refractive index of SiO_2 glass," *J. Non. Cryst. Solids*, vol. 135, no. 1, pp. 86–89, 1991.

[39] D. J. Monk, D. S. Soane, and R. T. Howe, "Determination of the Etching Kinetics for the Hydrofluoric Acid/Silicon Dioxide System," *J. Electrochem. Soc.*, vol. 140, no. 8, pp. 2339–2346, 1993.

[40] V. S. Ilchenko and A. B. Matsko, "Optical resonators with whispering-gallery modes—part II: applications," *J. Sel. Top. Quantum Electron.*, vol. 12, no. 1, pp. 15–32, 2006.

[41] F. Vollmer and S. Arnold, "Whispering-gallery-mode biosensing: label-free detection down to single molecules," *Nat. Methods*, vol. 5, no. 7, pp. 591–596, 2008.

[42] L.-M. Duan and H. Kimble, "Scalable Photonic Quantum Computation through Cavity-Assisted Interactions," *Phys. Rev. Lett.*, vol. 92, no. 12, p. 127902, 2004.

[43] S. I. Schmid, K. Xia, and J. Evers, "Pathway interference in a loop array of three coupled microresonators," *Phys. Rev. A*, vol. 84, no. 1, p. 013808, 2011.

[44] E. T. Jaynes and F. W. Cummings, "Comparison of quantum and semiclassical radiation theories with application to the beam maser," in *Proceedings of IEEE*, 1963, vol. 51, no. 1, pp. 89–109.

[45] S. M. Spillane, T. J. Kippenberg, K. J. Vahala, K. W. Goh, E. Wilcut, and H. J. Kimble, "Ultrahigh-Q toroidal microresonators for cavity quantum electrodynamics," *Phys. Rev. A*, vol. 71, no. 1, p. 013817, 2005.

[46] A. Beveratos, R. Brouri, T. Gacoin, A. Villing, J.-P. Poizat, and P. Grangier, "Single Photon Quantum Cryptography," *Phys. Rev. Lett.*, vol. 89, no. 18, p. 187901, 2002.

[47] J. Wolters, A. W. Schell, G. Kewes, N. Nüsse, M. Schoengen, H. Döscher, T. Hannappel, B. Löchel, M. Barth, and O. Benson, "Enhancement of the zero phonon line emission from a single nitrogen vacancy center in a nanodiamond via coupling to a photonic crystal cavity," *Appl. Phys. Lett.*, vol. 97, no. 14, p. 141108, 2010.

[48] F. Jelezko and J. Wrachtrup, "Single defect centres in diamond: A review," *phys. status solidi (a)*, vol. 203, no. 13, pp. 3207–3225, 2006.

[49] A. Lenef, S. Brown, D. Redman, S. Rand, J. Shigley, and E. Fritsch, "Electronic structure of the N-V center in diamond: Experiments," *Phys. Rev. B*, vol. 53, no. 20, pp. 13427–13440, 1996.

[50] Y. Mita, "Change of absorption spectra in type-Ib diamond with heavy neutron irradiation," *Phys. Rev. B*, vol. 53, no. 17, pp. 11360–11364, 1996.

[51] T. Gaebel, M. Domhan, C. Wittmann, I. Popa, F. Jelezko, J. Rabeau, A. Greentree, S. Prawer, E. Trajkov, P. R. Hemmer, and J. Wrachtrup, "Photochromism in single nitrogen-vacancy defect in diamond," *Appl. Phys. B*, vol. 82, no. 2, pp. 243–246, 2005.

[52] N. Manson, J. Harrison, and M. Sellars, "Nitrogen-vacancy center in diamond: Model of the electronic structure and associated dynamics," *Phys. Rev. B*, vol. 74, no. 10, p. 104303, 2006.

[53] F. Jelezko, C. Tietz, A. Gruber, I. Popa, A. Nizovtsev, S. Kilin, and J. Wrachtrup, "Spectroscopy of Single N-V Centers in Diamond," *Single Mol.*, vol. 2, no. 4, pp. 255–260, 2001.

[54] R. Brouri, A. Beveratos, J.-P. Poizat, and P. Grangier, "Photon antibunching in the fluorescence of individual color centers in diamond," *Opt. Lett.*, vol. 25, no. 17, pp. 1294–1296, 2000.

[55] M. Gregor, "Fiber Taper-Coupled Microresonators for Applications in Sensing and Quantum Optics," Humboldt-Universität zu Berlin, 2011.

[56] P. E. Barclay, C. Santori, K.-M. Fu, R. G. Beausoleil, and O. Painter, "Coherent interference effects in a nano-assembled diamond NV center cavity-QED system," *Opt. Express*, vol. 17, no. 10, pp. 8081–8097, 2009.

[57] Y. Y. Hui, Y.-R. Chang, T.-S. Lim, H.-Y. Lee, W. Fann, and H.-C. Chang, "Quantifying the number of color centers in single fluorescent nanodiamonds by photon correlation spectroscopy and Monte Carlo simulation," *Appl. Phys. Lett.*, vol. 94, no. 1, p. 013104, 2009.

[58] Y. Y. Hui, Y.-R. Chang, T.-S. Lim, H.-Y. Lee, W. Fann, and H.-C. Chang, "Erratum: 'Quantifying the number of color centers in single fluorescent nanodiamonds by photon correlation spectroscopy and Monte Carlo simulation' [Appl. Phys. Lett. 94, 013104 (2009)]," *Appl. Phys. Lett.*, vol. 94, no. 14, p. 149901, 2009.

[59] J. Michaelis, C. Hettich, J. Mlynek, and V. Sandoghdar, "Optical microscopy using a single-molecule light source," *Nature*, vol. 405, no. 6784, pp. 325–328, 2000.

Chapter 5

Silica-Based Waveguide Structures

5.1 Introduction

Within all fields of modern optical technologies, a strong trend towards higher levels of integration can be observed. When compared to conventional systems based on discrete optical elements, photonic integrated circuits (PICs) allow the realization of compact optical systems containing hundreds of components. PICs often show highly increased performances and enhanced system stability mainly due to their compact sizes, and high levels of production and function reliability. The optical functionality of PICs can be further enhanced by combining the photonic elements with additional electronic components. As for conventional electronic integrated circuits (EICs), highest levels of integration and the wide scalability of all involved production processes are important aspects to become a major technological breakthrough. PICs are already used within the field of optical communication [1], for optical interconnects [2], and in sensing applications [3].

In the field of quantum optics, PIC-based lab-on-chip designs offer great advantages over standard table-top experiments. When dealing with single photons, optical losses are the most crucial parameter. The conventional experimental setups in quantum optics often consist of large amounts of optical components. By integrating most of these devices onto a single photonic microchip, the number of interfaces can be highly reduced and optical losses can be minimized. Due to the strong reduction in the experimental volume, the influence of external mechanical or electronical noise is also reduced and a large enhancement in the overall system stability can be achieved. However, integrating classical photonic elements, e.g., splitters, ring resonators, and interferometers, onto a single optical microchip requires a highly efficient waveguide technology which allows interconnecting all relevant parts of such a PIC with lowest transmission and insertion losses.

The design and manufacturing processes for such devices highly depend on the used material platform. Over the last decades, several waveguide compatible systems were investigated. For applications mainly working in the short-wavelength infrared (IR-B) telecommunication range, silicon nitride (Si_3N_4) is widely used [4], [5]. However, for most telecommunication applications, silicon-on-insulator (SOI) is the platform of choice [6], [7]. In so-called silicon photonics, this material system allows highest levels of integration and standard EIC processing schemes can be applied. SOI wafers were originally designed for electronic device manufacturing. The high purity of this material system allows exceptional optical properties to be observed. This makes SOI wafers an ideal material basis for integrated photonics in the IR-B spectral range. By changing the dielectric material of the waveguide, the applications can be extended into other spectral ranges. For the visible (VIS) spectral range, gallium phosphide (GaP) is a common material [8]. Other material systems are based on $LiNbO_3$ [9] or III-V semiconductors [10]. For working in the ultraviolet (UV) spectral range, different types of polymers can be used [11].

In the field of quantum optics, the most crucial parameters are low transmission losses, low background florescence, and a broad transmission window. Hence, for working in the VIS spectral range, silica is the perfect waveguide material [12], [13]. By thermally oxidizing standard silicon wafers, highly pure layers of silica with exceptional optical properties can be produced. The high purity of this thermally oxidized silica also allows extending the range of possible applications into the UV spectral range. The main problem for using this material in photonics is the relatively low refractive index of silica. It prevents using many of the conventional waveguide designs which often require to be directly structured on top of a substrate. For supporting the waveguides, a material with lower refractive index has to be used. With silicon or other dielectrics having a higher index of refraction, the guided optical modes are coupled into the substrate and confined mode propagation can not be achieved. Furthermore, additional material absorption within the substrate will increase the overall transmission losses. Both effects are strongly damping the intensity of the transmitted light. They are the main reasons for the high losses commonly observed within such directly stacked waveguide structures.

Within this chapter, a novel waveguide design for silica-based material systems is presented. The resulting transmission properties of the structures are analytically and numerically investigated and the corresponding fabrication scheme is shown. With this approach, volume processing of highly pure silica waveguides becomes possible on chip scale. Standard wafer processing techniques can be applied for fabricating the desired optical waveguide structures. Although the given design approach aims for the microintegration of quantum optics experiments, the method is also suitable for a wide range of other applications requiring low-loss air-cladded waveguide designs in silica and other suitable material systems. The presented design was developed and realized in close cooperation with the Ferdinand-Braun-Institut, Leibnitz-Institut für Höchstfrequenztechnik (FBH). The carried out works resulted in a collaborative patent application [14].

5.2 Structures

In general terms, the aimed problem can be described as designing a planar waveguide structure with an index of refraction smaller than the index of refraction in a supporting substrate. In practice, three different techniques are known for solving this problem.

The first approach is based on deeply etched waveguide structures (see FIG. 5.1a). For separating the waveguides from the substrate, a core layer with low refractive index contrast is structured within a homogeneous cladding material. The stack is usually processed with extended etching steps such that deep trenches are formed. They often reach down to the substrate by etching 16 μm or more into the superstrate. For low-loss optical waveguides, a high surface quality of the sidewalls is important. In deeply etched structures, this can only be achieved when advanced processing techniques are applied. The deep etching method is compatible with arbitrary material systems. In silica, the realization of deeply etched optical waveguides and other devices has already been demonstrated [15]. The side walls of the resulting structures are fully air-cladded. This allows designing optical circuits with ultra-compact sizes. Bending losses can be neglected and smallest waveguide bend radii can be realized [16]. Another advantage of air-cladded designs is the possibility to access the evanescent fields of the guided modes. This can be used in sensing applications or for evanescently coupling optical fields to other photonic structures, e.g., resonators or other waveguides. However, the fabrication is difficult and several repeated etching steps are required. The produced sidewalls are often very rough and the guided optical fields are intensively scattered on these defects. Furthermore, the modes are not fully confined within such structures and high transmission losses are commonly observed for such waveguide designs [17].

Another option is to place the waveguides directly on top of a structured multilayer. It consists of several alternatively ordered high and low refractive index materials. The structure acts as antiresonant Bragg reflection grating underneath the waveguides. It prevents the optical modes from leaking into the substrate. The grating works as a substrate mirror which confines light to the waveguide. These so-called antiresonant reflecting optical waveguides (ARROWs) can be used in various material systems [18].

The stripe-line pedestal anti-resonant reflecting optical waveguide (SPARROW) is an example for implementing this method in silica-based technology (see FIG. 5.1b) [19]. However, also this technique shows some drawbacks. One is the high technological effort involved in producing the multilayered structures on top of an arbitrary substrate. Furthermore, the silica for the waveguide has to be deposited and can not directly be oxidized from a silicon substrate. Another drawback is the requirement of designing an appropriate layer structure to suppress mode leakage. It must be specifically engineered for the applied wavelength range and is often a tradeoff between low leakage and high transmission bandwidth. Well designed structures show nearly vanishing transmission losses, but work only in a limited wavelength range. Other designs may have a broader bandwidth, but show very high substrate leakage. Furthermore, at least some part of the evanescent field of the guided modes always interacts with the layered structure. This introduces additional transmission losses compared to simple waveguiding in highly pure silica. This can be critical especially for applications working in the UV spectral range. In quantum optics, even the interaction between single photons and included dopants can be decisive. Background fluorescence from this interaction often strongly disturbs the measured signals.

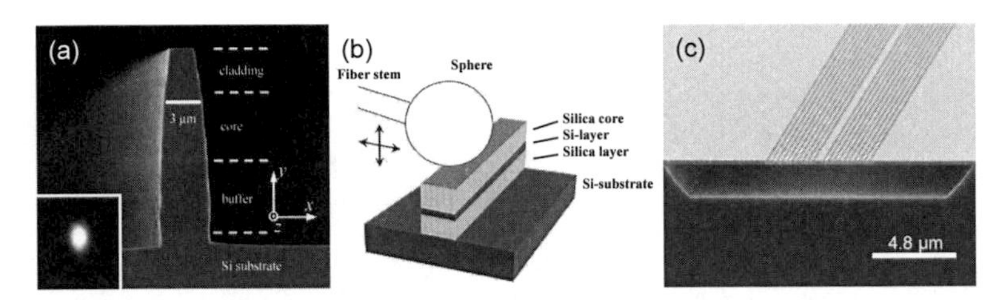

FIG. 5.1. (a) Scanning electron micrograph of the cross-section of a deeply etched silica waveguide (from [15]). The inset shows the corresponding mode profile as seen by a CCD camera. (b) Schematic of an integrated stripe-line pedestal anti-resonant reflecting optical waveguide coupler for microsphere WGM excitation. The reflector is build of a single layer of silicon (from [19]). (c) Scanning electron micrograph of an air-cladded GaAs-based single line-defect two-dimensional photonic crystal slab waveguide (from [20]).

Another widely used technique to solve the problem of separating a low refractive index waveguide from a high refractive index substrate is by using two-dimensional (2D) photonic crystal structures [20]. Here, in contrast to the previous designs, the waveguides are completely separated from the underlying substrate (see FIG. 5.1c). By underetching the guiding layer, a fully suspended membrane is produced. In this case, substrate leakage can be avoided and waveguiding is performed within this membrane. Lateral confinement of guided modes is carried out by the photonic band gap effect. The functionality of such optical systems can be controlled by precisely structuring the local environment of the waveguides. This technique can also be applied to a variety of other material systems and nearly arbitrary distributions of the involved refractive indices. By using the process of generating photonic elements by refractive ion etching (GOPHER), it is even possible to produce this 2D photonic crystal structures in homogeneous materials [21]. The process was originally developed for thick silicone membranes, but can also be adopted for silica and other material systems.

The main drawbacks of this waveguide technology are the high technological efforts during the production processes and strong demands on the precise definition of the intended crystal structure. While in classical waveguide designs the transmission losses mainly depend on the sidewall surface roughness, in 2D photonic crystals the whole volume structure is responsible for creating the effect of waveguiding. Roughness also plays an important role, but more important are geometrical aspects of the involved elements. Uniformity and the exact periodicity of the produced structures have to be ensured. Even smallest deviations from the optimal design can break the symmetry of the system and lead to enhanced optical losses. All relevant structure sizes are well below the operational wavelength of the devices. This places high demands on the applied structuring processes and prevents mass production. The whole technological process is still matter of research, but even with modest processing techniques, low-loss 2D photonic crystal waveguides are limited to lengths of around hundred wavelengths at maximum. Photonic crystals can not be used to bridge large distances as they are commonly appearing on photonic microchips.

All three designs are not suitable for the realization of long ranging waveguides in highly pure silica. For compact photonic microchip designs and applications requiring ultra-low optical losses, a suitable production scheme is still missing. However, for lab-on-chip applications in quantum optics, silica-based waveguides are perfectly suited due to the exceptional optical properties of the material. For overcoming the discussed problems, an alternative waveguide design was developed.

5.3 Design

In the following sections, the discussion of the novel waveguide design is mainly concentrated on silica-on-silicon as basic material. The design itself is not limited to this material, but the developed fabrication method belongs specifically to that system. For other material systems, alternative processing techniques from the semiconductor industry can be applied.

For developing a proper silica waveguide structure for the intended quantum optics applications, specific design constraints had to be considered. The waveguides must be optimized for conditions working with ultra-low light intensities at the single photon level. In such applications, optical losses are the most crucial factor. Beside a strong dependency on the intrinsic properties of the used material, especially optical losses caused by the fabrication processes or induced by coupling the waveguides to glass fibers or via free space beams must be also considered.

For producing the waveguide structures, a very simple fabrication scheme mainly using standard lithographic structuring and etching processes is preferred. Allowing for mass production, mask-based structuring methods are thus highly favored over direct e-beam lithography. For maintaining the high purity of the basic material, solely removal steps were allowed for structuring. In principle, the deposition of silica layers would also have been possible by using chemical vapor deposition (CVD) or other techniques. However, the final optical quality of such deposited materials strongly depends on the applied processing conditions and high technological efforts are involved [22].

Another key design constraint was the mechanical stability of the produced structures. Elements with length up to the cm-range are required. For practical applications, an efficient method for coupling to fibers or free beams is a further requirement. As planar face-to-face butt-coupling is most efficient, the facets of the structured waveguides should be accessible from both ends. This increases the demands for system stability as the separation between substrate and waveguide has to be guarantied over the full length of the waveguide element. If the separation of the guiding regions is achieved by underetching, the intrinsic material stability is widely lost and the waveguides have to be stabilized by an additional support.

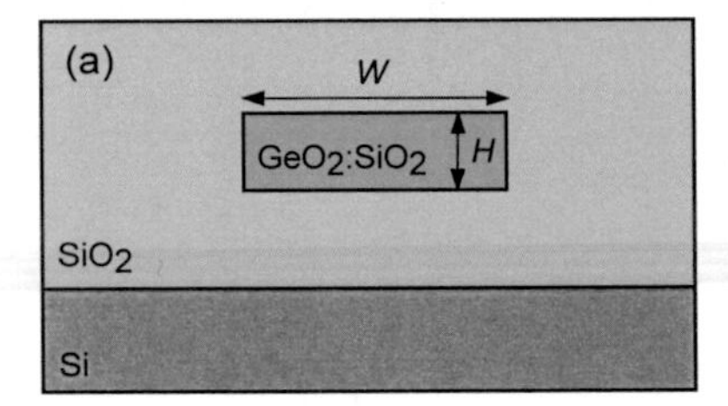

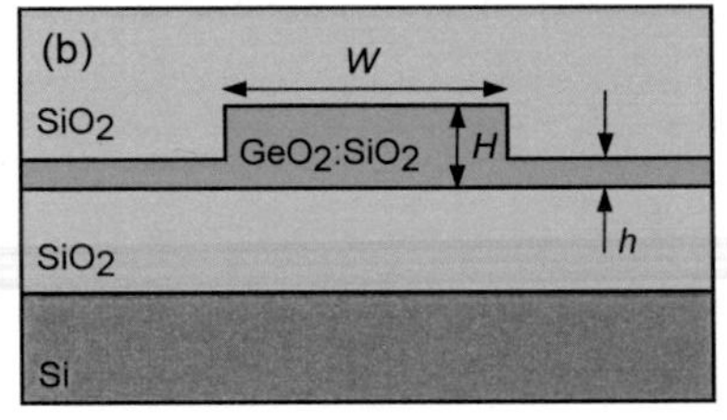

FIG. 5.2. Different types of optical waveguides. A silica layer is deposited or grown on top of a silicon substrate. The waveguide core is structured by doping some parts of the silica with germanium oxide to locally increase the refractive index of the material. (a) Buried channel waveguide. (b) Rib waveguide.

In a first step, an appropriate waveguide structure was chosen. For planar chip-based applications, several types of dielectric waveguide designs can be used. Most common are different types of slab [23], buried channel [24], stripe [25], rib [26], and ridge [27] waveguides. The main difference is how the optical mode is confined to the core region of the waveguide elements. While in a slab waveguide light is confined within a homogeneous core material with a high index of refraction only in one dimension (1D), i.e., between the upper and lower side of the waveguide, the buried channel waveguide (see FIG. 5.2a) is completely capped within a homogeneous lower index cladding material in two dimensions.

A standard rib waveguide (see FIG. 5.2b) can be described as a combination of a slab and a buried channel waveguide. Depending on the height H of the rib and the wavelength λ of the guided light, the propagating mode is trapped inside a potential around the ridge region by a high refractive index contrast between waveguide core and cladding. The mode is confined over long distances even there is a small residual slab at the bottom of the guiding region. A major disadvantage of such waveguide designs is the spatial distortion of the intensity distribution. In a buried channel waveguide, a nearly point symmetric intensity profile allows excellent coupling to the Gaussian modes of an optical fiber or to free beams. In a rib waveguide, this profile may be widely stretched along the direction of the slab. This will decrease the achievable coupling efficiency between fiber and waveguide modes as their spatial overlap is the most crucial parameter here. However, it is possible to reduce the thickness h of the slab regions such that no propagational modes are allowed within these volumes. In this case, the properties of a ridge waveguide are comparable to a stripe waveguide and symmetric mode profiles can be achieved even with residual slab material located at the sides of the rib.

A rib waveguide design with suppressed mode propagation within the slab regions is in perfect agreement with the intended applications. Thus, a corresponding silica-based design was developed. It allows underetching the intended waveguide structure for separating the optical active regions from the underlying substrate. The final waveguide design is presented in FIG. 5.3.

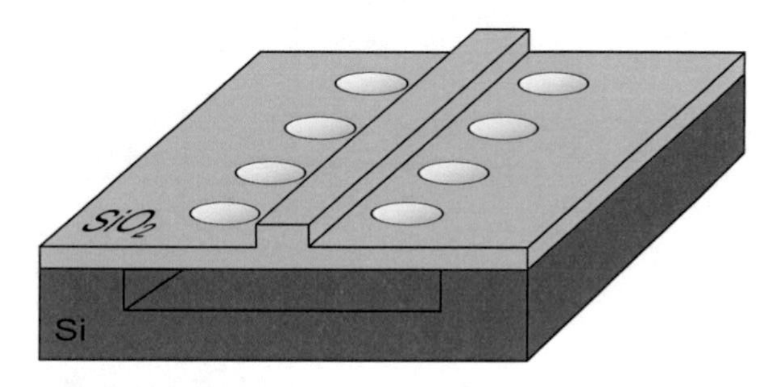

FIG. 5.3. Schematic view of the developed air-cladded waveguide design. The waveguide and the supports are made from silica, the substrate material is silicon. The waveguide section is fully suspended and separated from the substrate by partially underetching the whole structure.

The rib waveguide is suspended from two sides by thinned silica slabs. They bridge the gap between the supporting side areas by being in direct contact with the silicon substrate. Etching holes in the silica allow a selective isotropic etching of the silicon directly underneath the waveguide. The holes are positioned in a certain distance from the waveguide. Their final spacing must allow complete separation between waveguide and substrate, but still ensure a sufficient mechanical stability of the whole structure. On the other hand, with decreased spacing, the horizontal extend of propagating modes can interfere with the holes and additional transmission losses may occur. Hence, the mode structure of the waveguide has to be considered and proper placement of etching holes must be guaranteed to avoid such effects.

5.4 Analysis

The presented silica waveguide design can be classified as a rectangular dielectric waveguide. This type of a planar structure is standard in integrated photonics and many different techniques and methods were developed to analytically and numerically describe its internal modes. However, it is not possible to derive a general analytic solution for these structures and approximations have to be applied to analyze the problem [28].

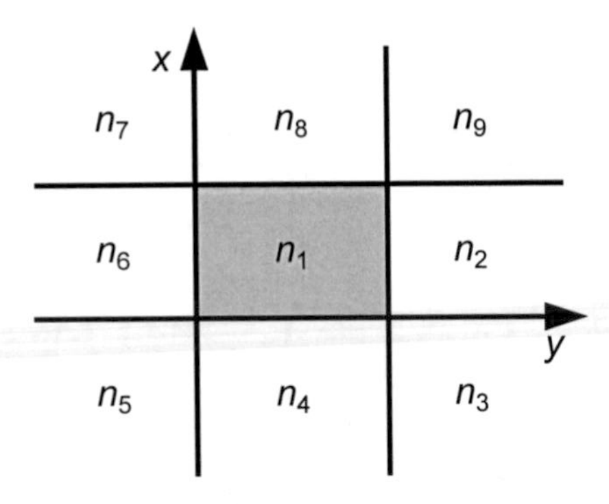

FIG. 5.4. General model for the analytic description of a rectangular waveguide. The middle section with refractive index n_1 is the core. The surrounding areas are required to have lower refractive indices $n_i < n_1$.

For an arbitrary rectangular waveguide, the 2D cross-section of the structure can be divided into nine different sections as shown in FIG. 5.4. The waveguide core is located in the middle section; it normally shows the highest index of refraction. For an exact analytic solution of the problem, all involved boundary conditions must be satisfied. At the four corner regions, the x and y solutions of the different fields are coupled to each other. The resulting wave equation can not be solved with the standard method of variables separation and full analytic solutions can not be calculated. However, if the guided electromagnetic modes are tightly confined to the core region, the influence of the four corner regions may be negligible and variables separation can still be applied.

A buried channel waveguide is fully embedded within a homogeneous material with lower refractive index. For analyzing such a design, the Marcatili approximation is a suitable solving approach [29]. With this method, the refractive index in the four corner regions is artificially reduced and the geometry is separated into two orthogonal slab waveguides. In this case, the one-dimensional slabs can be independently solved and a full solution can be found. Furthermore, the accuracy of the result can be increased by applying a perturbative correction to the simple solution obtained with the Marcatili approximation [30]. With such an approach, also the influence of the neglected fields in the four corner regions can partly be considered.

5.4.1 Effective Index Method

For calculating the given silica waveguide design, another approach is more suitable. The full structure can be sub-classified as a special type of symmetric rib waveguide. For analytically solving such kinds of waveguides, often the well-known effective index method (EIM) is applied [31]. The Marcatili approximation normally involves extensive analytic calculations which can be avoided by using the EIM. The 2D waveguide problem is here again separated in two 1D problems, but in contrast to the Marcatili approximation, the two separated problems are not independently solved. Instead, the interaction of the two separate fields is fully taken into account. After solving one dimension of the problem, the second dimension is solved as perturbation of the first one. The EIM is ideally suited for rib waveguides as in this case the problem effectively reduces to the solution of two simple slab waveguides. For this method, a variety of further correction methods can also be applied to improve the accuracy of the calculated results [32].

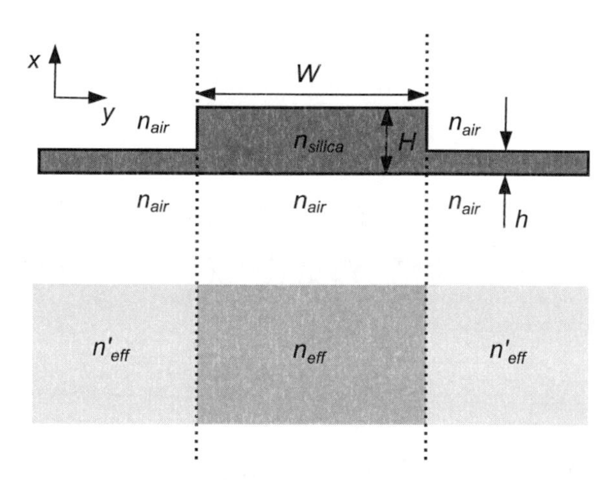

FIG. 5.5. Effective index approximation for the given silica waveguide design. The waveguide is completely air-cladded by all sides. In a first step, the effective index of the vertical sections is calculated. With these results, an equivalent horizontal slab waveguide is calculated in a second step.

In the following section, the EIM is used to calculate the size and shape of the fundamental mode in the presented rib waveguide design. The analytic results are compared to results of numerical simulations. Further analysis allows estimating the maximum allowed thickness of the supporting side structures before the guided optical modes begin to leak out into these areas. The lateral extend of the modes is also calculated as it defines the minimum distance between waveguides and etching holes.

The design wavelength for the intended silica waveguides is $\lambda = 780$ nm. It is assumed that the guided mode is a TM mode and thus polarized along the vertical x-axis of the waveguide. The refractive index of silica is $n_{silica} = 1.45367$ for the given design wavelength [33]. The refractive index of air is considered with $n_{air} = 1$. For the guiding silica rib section, a height $H = 2$ µm is assumed; the thickness of the surrounding side structures is set to $h = 200$ nm. For the rib section, a waveguide width of $W = 4$ µm is selected. The assumed dimensions are based on technological requirements and are tradeoffs between achievable optical properties and fabrication constraints.

For applying the EIM, the structure is separated into three different horizontal sections. As the waveguide structure is completely surrounded by air, all these sections can be considered as simple symmetric slab waveguides. A schematic of the presented design can be also found in FIG. 5.5.

The characteristic equation for TM modes in such a symmetric slab is given by [28]

$$\tan(\kappa_x \cdot \frac{H}{2}) = \frac{n_{silica}^2}{n_{air}^2} \cdot \frac{\gamma_x}{\kappa_x} \tag{5.1}$$

with the attenuation coefficient $\gamma_x = \sqrt{k_0^2 (n_{silica}^2 - n_{air}^2) - \kappa_x{}^2}$. κ_x is the transverse wave vector component in the x-direction. The vacuum wave number is given by k_0.

By using $\beta_x = \sqrt{k_0^2 \cdot n_{silica}^2 - \kappa_x^2}$ for the longitudinal wave vector component, Eq. (5.1) can be solved for the rib section of the structure. The effective refractive index of this section is then calculated as

$$n_{eff,x} = \frac{\beta_x}{\kappa_x} = 1.442 \,. \tag{5.2}$$

From a corresponding calculation for the effective refractive index of the two side regions, it is found as $n_{eff,x}' = 1.093$.

The three effective refractive indices from the vertical sections can be used to calculate the effective waveguide in the horizontal direction as it is shown in FIG. 5.5. In this direction, the x-polarized wave appears as a TE mode. The characteristic equation of this configuration is given by [28]

$$\tan(\kappa_y \cdot \frac{W}{2}) = \frac{\gamma_y}{\kappa_y} \tag{5.3}$$

with the attenuation coefficient γ_y and the transverse wave vector component κ_y in y-direction. This equation can again directly be solved and an effective refractive index of $n_{eff,y} = 1.439$ is calculated. Thereby, the calculated wave vector component β_y is also the eigenvalue of the corresponding waveguide mode.

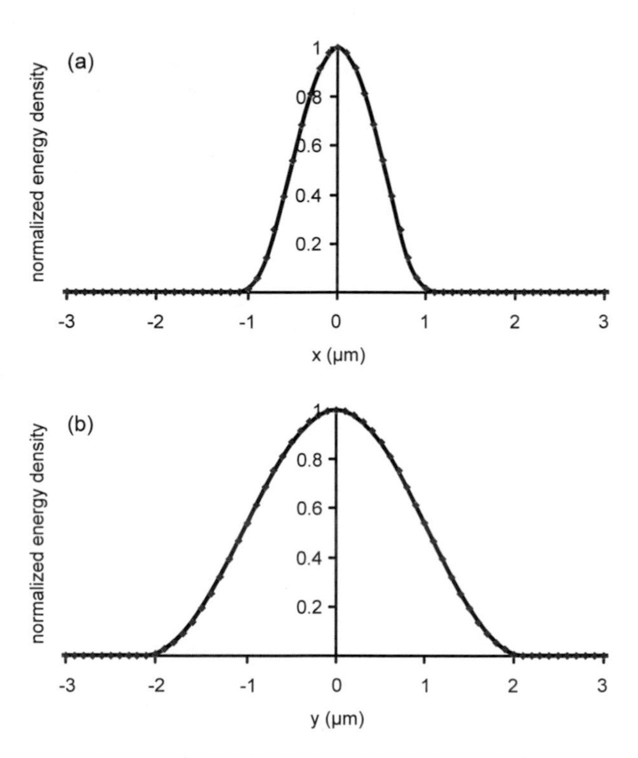

FIG. 5.6. Normalized energy density distribution of the fundamental TM mode in air-cladded silica waveguides along the (a) x- and (b) y-direction. A rib size of 2 x 4 µm was assumed. The thickness of the two side structures was 200 nm, the design wavelength was 780 nm, and the silica refractive index at this wavelength was assumed with $n_{silica} = 1.45367$. The dotted lines are results obtained from numerical simulations of the same microstructure.

The calculated effective waveguide is symmetric. In this case, the analytic description of the fundamental mode in the waveguide is given by [28]

$$E(x)=\begin{cases} A\cdot\dfrac{\cos(\kappa_x\cdot x)}{\cos(\kappa_x\cdot H/2)} & |x|<H/2 \\ B\cdot e^{-\gamma_x|x|} & |x|>H/2 \end{cases} \tag{5.4}$$

and

$$E(y)=\begin{cases} C\cdot\dfrac{\cos(\kappa_y\cdot y)}{\cos(\kappa_y\cdot W/2)} & |y|<W/2 \\ D\cdot e^{-\gamma_y|y|} & |y|>W/2 \end{cases}. \tag{5.5}$$

With the previously calculated propagation parameters, these equations can be fully solved. The resulting mode profiles of the fundamental mode in x- and y-direction are shown in FIG. 5.6.

The results found by the EIM (black lines) are compared to results obtained from a numerical simulation of the same problem (dotted line). This simulation was performed with a commercial full vectorial eigenmode solver which was also used for calculating optical modes in microresonators (Photon Design Fimmwave, see also Appendix A). The two different mode distributions agree well with each other and proof the validity of the EIM approach for the given design. For the initially used size parameters, the horizontal and vertical extents of the field are corresponding to the geometric parameters of the rib. This is due to a relatively high confinement of the optical mode. It allows placing the etching holes in nearest proximity to the waveguide structures without inducing additional scattering losses. For the given size parameters, a minimum distance of 1 µm can be estimated as minimum lower bound for a separation between waveguides and etching holes.

5.4.2 Numerical Aperture

The numerical aperture of a waveguide is defined by the maximum angle of incidence under which a ray of light can by guided within the waveguide. In the ray optics picture, this angle is given by rays that barely satisfy the condition of total internal reflection on the interface between waveguide and cladding material. The numerical aperture can be defined by the sine of the acceptance angle of the waveguide as

$$NA=\sin(\vartheta_{\max})=\sqrt{n_{core}^2-n_{cladding}^2}\,. \tag{5.6}$$

In case of the presented structure, the waveguide is cladded by air and the conditions for strong guiding are fulfilled. At a wavelength of λ = 780 nm, the refractive index of silica is n_{core} = 1.45367 [33]. The refractive index of air is approximated by $n_{cladding}$ = 1.

With Eq. 5.6, a numerical aperture $NA = 1.056$ is calculated from these values. There is no limitation for the acceptance angle and any light that is incident to the surface can be guided. However, this also causes an extremely wide output cone for rays guided under large angles. Thus, the efficient collection of higher order waveguide modes is difficult with such a design.

5.4.3 Mode Profiles and Far-Field Properties

For a rigorous analysis of the propagation constants and estimating the mode profiles within the waveguide, additional numerical simulations were performed with the used numerical simulation software (Photon Design Fimmwave). The complete waveguide design was modeled within the internal CAD module of the program. All intrinsic material properties and the boundary conditions were implemented according to the specific structural design defined in the beginning of Section 5.4.1. For the simulations, the internal 2D finite element method (FEM) solver of the software was used. This allows a full vectorial analysis of all relevant waveguide modes.

The results of these simulations are presented in FIG. 5.7a. Thereby, the TM profiles of the fundamental and the first three higher order modes are shown. The geometric parameters for the simulated waveguide correspond to the values assumed for the analytic EIM solution shown in Section 5.4.1. All mode profiles are fully bound to the waveguide. Even for the highest order mode, there is no energy leakage into the side regions observed. The initial geometry already fulfills the requirements of the desired waveguide design and is thus suitable for developing a corresponding manufacturing process.

With information about the exact field distribution inside the waveguide, the far-field profiles of the different modes can be calculated. A far-field plot describes the radiation properties of a single optical mode exciting the waveguide and then propagating in free space. This is an important aid to estimate the achievable coupling efficiency between a specific waveguide mode and modes of an optical fiber or other external coupling devices. The far-field properties can directly be calculated by the applied simulation software. A plane wave expansion is used to decompose the near field at the output facet to find an angular distribution for the free-space propagation. This expansion is defined by

$$E = \sum_m A_m \cdot e^{i \cdot 2\pi \cdot k_x \cdot x} e^{i \cdot 2\pi \cdot k_y \cdot y} = \sum_m A_m \cdot e^{i \cdot 2\pi \cdot \frac{\cos(\vartheta_x)}{\lambda} \cdot x} e^{i \cdot 2\pi \cdot \frac{\cos(\vartheta_y)}{\lambda} \cdot y} \tag{5.7}$$

with A as the amplitude of a plane wave exciting the structure at angles ϑ_x and ϑ_y into the direction of free-space propagation. The software uses an internal FFT-based rigorous vectorial algorithm to perform the calculations. This allows the full vectorial analysis of the far-field emission properties.

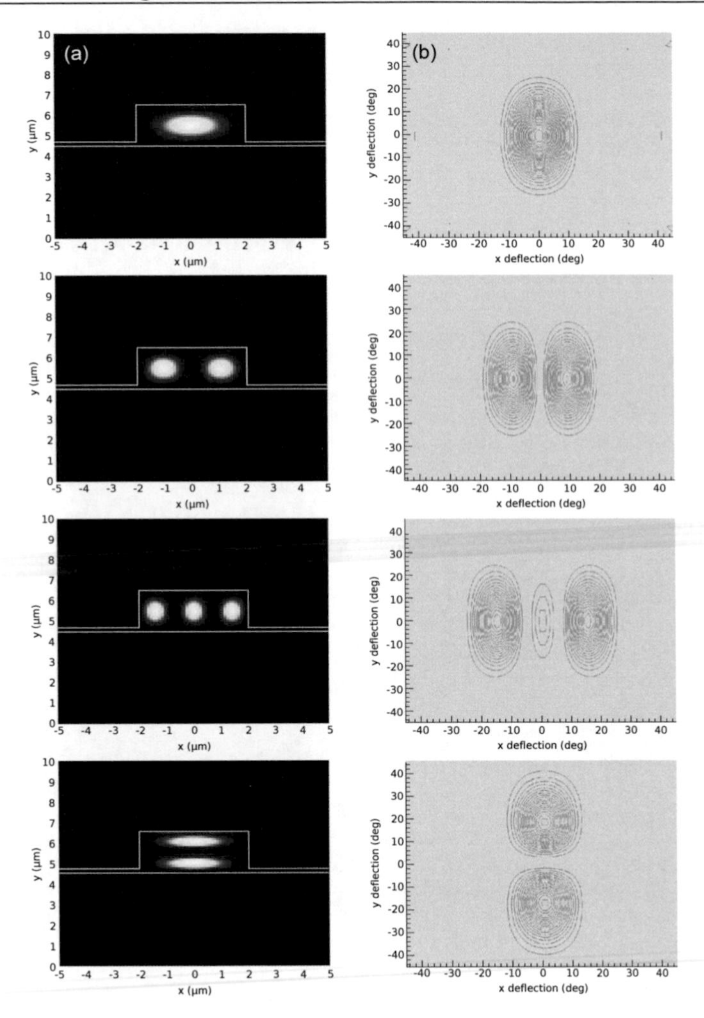

FIG. 5.7. (a) Mode profiles numerically calculated with Fimmwave (λ = 780 nm, n_{air} = 1, n_{silica} = 1.45367, W = 4 µm, H = 2 µm, h = 200 nm). (b) Far-field radiation properties of the corresponding waveguide modes.

The corresponding far-field profiles of the calculated modes are shown in FIG. 5.7b. For the fundamental TM mode, a full divergence angle (1/e^2 width) around ϑ_x = 20.2°

was calculated in the horizontal direction. In the vertical direction, an angle of around $\vartheta_y = 38.8°$ was observed. For the first three higher order modes, the cone angle nearly doubles in the horizontal direction. In the vertical direction, no influence on the cone angle could be observed for the higher order modes.

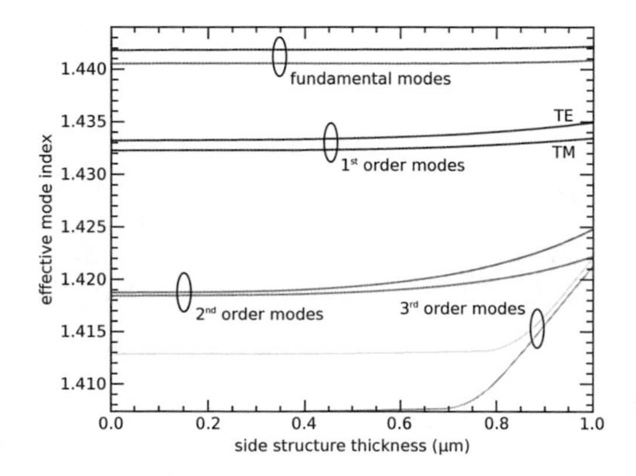

FIG. 5.8. Dependence of the effective indices of the first four mode orders (TE and TM) on the thickness *h* of the waveguide side structures.

Numerical simulations allow analyzing the influence of the support areas with highest accuracy. Hence, the propagation constants of different modes were calculated in relation to the thickness of the side structures. The results of these simulations are shown in FIG. 5.8. For the presented waveguide geometry, the side structure thickness *h* was varied between 0 µm and 1 µm. This corresponds to thicknesses between 0 % and 50 % of the waveguide height *H*. For the lowest order modes, no influence could be observed. For the higher order modes, the thickness of the side structures becomes more critical. A strong thickness dependence of the modes is observed for values starting at around 25 % of the waveguide height.

5.5 Fabrication

The presented waveguide design was specifically developed for silica (SiO_2) guiding layers thermally grown directly on silicon substrates. Beside the good optical properties of such kind of silica, in Si/SiO_2 material systems well-known structuring techniques from semiconductor industry can be applied. However, the presented design can also be used for other material systems.

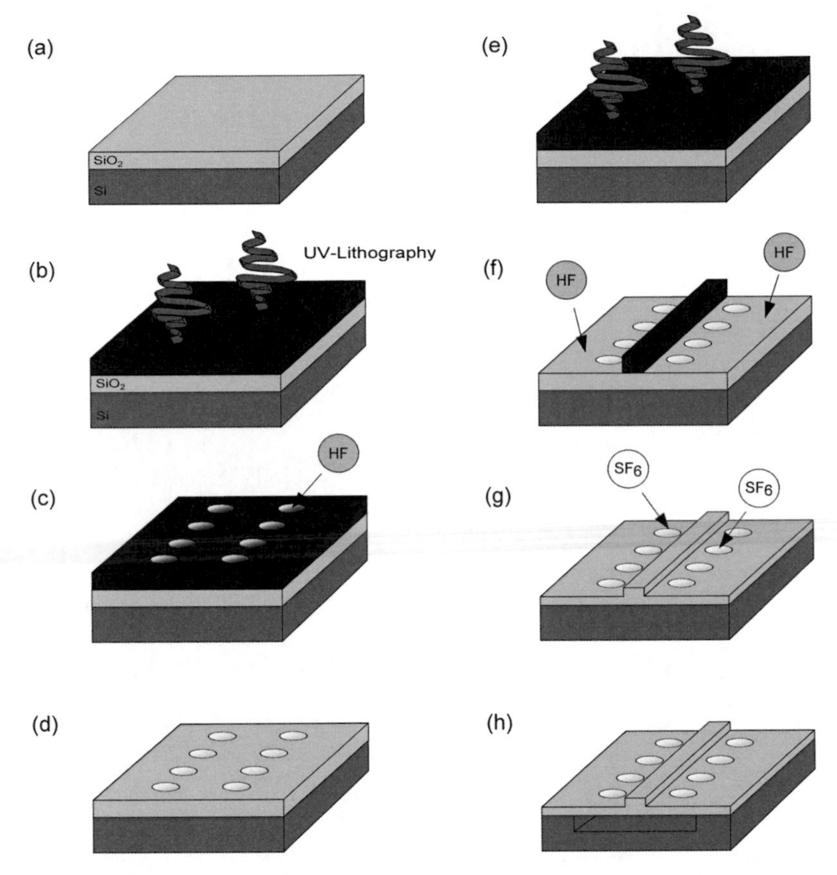

FIG. 5.9. Fabrication process for air-cladded silica waveguides. Silica layer etching is carried out by wet chemical processing with buffered hydrofluoric acid (BHF). For underetching the silicon substrate, dry plasma etching (PE) with sulfur hexafluoride (SF_6) is applied. The details of the process are explained in Section 5.5.

The waveguide structures were produced from commercially available single crystalline standard electronic grade silicon substrates with thin layers of amorphous silica on top. The silica was grown into the substrates by means of a standard dry or wet chemical oxidation process. At temperatures around 1100 °C, silicon wafers are exposed to an oxygen-rich atmosphere or immersed into a water bath. Thereby, the thickness of the produced silica layers can be controlled by the overall processing time.

A thermal oxidation process is very slow (up to 24 hours per micrometer) and only thin layers can be produced by this method (up to a few micrometers). Thermally grown silica is known for its exceptional optical and mechanical properties as well as for a high material density. The oxidized wafers can be bought from various commercial manufacturers. For the presented waveguide design, wafers with a 2 µm thick silica layer were chosen.

It is also possible to deposit silica with various methods, e.g., by using plasma enhanced chemical vapor deposition (PECVD) [13] or flame hydrolysis deposition (FHD) [34]. The density of the deposited layers is much lower compared to thermally grown silica. The optical properties of such porous silica are also often reduced. However, silica can be used as waveguide material on nearly arbitrary substrates by using deposition techniques. This is highly interesting for the implementation of more elaborated photonic designs were heterogenic integration with active optical elements, e.g., lasers or detectors, is required.

The detailed fabrication process for the presented waveguide design is presented in FIG. 5.9. For structuring the different sections, two subsequent lithography and three different etching steps are involved. Mask-based lithography is used, thus the two-step process requires consecutive alignments between the two masks with high precision. By using special alignment markers, which are transferred to the wafer within a first etching step, automatic alignment to the second mask is possible with a mask aligner. The waveguide structuring process is designed to be fully compatible with standard contact lithography techniques and allows the automated mass production of devices.

In a first step, the wafers are carefully cleaned from organic and inorganic residues. Afterwards, a layer of negative photoresist is deposited on top of the wafers via spin coating. For allowing the separation of the waveguides from the silicon substrate by underetching, etching holes have to be placed along both sides of the waveguides. Hence, in a first lithography step, the etching holes are transferred to the resist by illuminating a corresponding photomask with ultraviolet light. Thereby, the resist is developed and the transferred structures are washed out by immersing the wafer into a solvent. The etching holes are created by partially etching the silica down to the substrate. Therefore, a buffered hydrofluoric acid (BHF) wet chemical etching solution is used. Also a dry chemical etching method, e.g., by using SF_6 or XeF_2 gases for reactive ion etching (RIE) or plasma etching (PE), can be applied.

After etching the holes and removing residual resist from the surface, a second layer of photoresist is deposited. Then a second lithography mask containing the waveguide structures is carefully aligned to the previously defined alignment markers. The waveguides are directly structured in between the rows of etching holes. After developing the resist and transferring the waveguides onto the wafer, the patterned structures are washed out again. Now a second etching step is applied to the silica layer. This process has to be finished before the underlying silicon is reached. Thus, the silica is partially thinned to create the supporting side structures for the waveguides.

Within this thinning process, rib waveguides as analyzed in the previous sections are produced. The residual thickness of the silica can be controlled by adjusting the overall etching time. The rates of a wet chemical BHF etching process are very reliable and can be controlled with highest precision. In the presented waveguide design, a residual layer thickness of 200 nm was chosen.

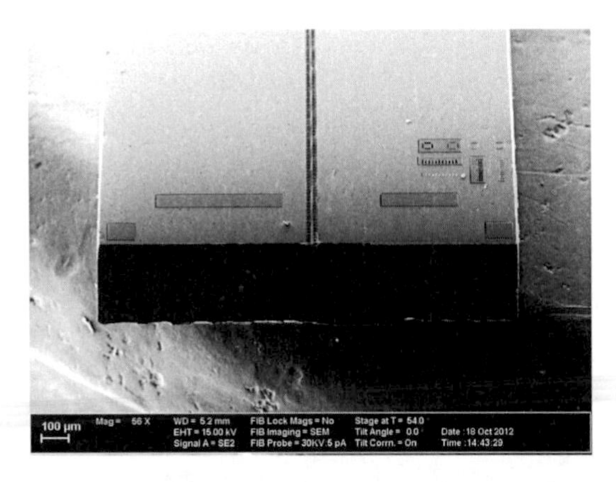

FIG. 5.10. Scanning electron micrograph of the presented silica-based waveguide structure directly after fabrication.

After the second etching, residual resist is again removed. The structured area can then be separated from the substrate. For that, in a third etching step, a highly isotropic dry PE process with an inductively coupled SF_6 plasma is used. This plasma has a high selectivity against silica and allows removing the underlying silicon through the structured etching holes. The process etches the silicon substrate in the horizontal and vertical direction with comparable rates. Thus, the waveguides are highly isotropically undercut. The process completely separates the waveguides from the substrate and the resulting ribs are fully suspended by thinned side structures around the etching holes.

For an additional stabilization of the waveguides, even though it will increase the overall transmission losses, the etching process can be stopped before the undercuts are completed. In this case, small silicon pillars will remain underneath the waveguide structures. The etching rates through the holes are highly reproducible. By adjusting the position and size of some of the etching holes, just a few pillars will remain while others are completely dissolved. With such a technique, a partial stabilization of the waveguide design can be achieved in regions which require higher rigidity, e.g., where a completely free standing waveguide may be used to evanescently couple light into a nearby microresonator.

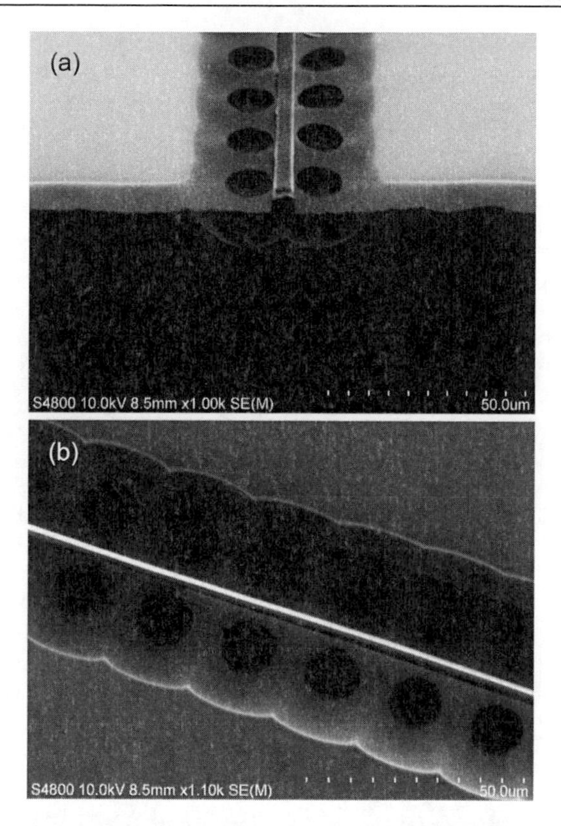

FIG. 5.11. Scanning electron micrograph of the produced silica-based waveguide structure. Underneath the top layer, the silicon substrate can be seen. (a) Cross-section of the structure showing the edge of a diced chip. The sides of the silicon are also etched such that a thin outer membrane is created. (b) Top-view of the rib waveguide and the produced etching holes.

For improving the result of the fabrication process, several thermal reflows are performed on photoresist and silica. Furthermore, a final etching dip in concentrated BHF can be used to chemically polish the resulting surfaces. By using such refinement methods, the resulting surface roughness of the structures can be considerably decreased.

After structuring different elements onto a single microchip, the waveguides have to be separated from each other. This so-called dicing process can be carried out by standard wafer sawing or by simply breaking out the devices from the wafer. For the presented waveguide design, these two types of direct dicing were not directly applicable due to the fragility of the supporting silica structures. Both dicing processes produce strong mechanical vibrations inside the silica layer which cause the suspending structures to collapse. Instead, the wafers had to be diced before performing the final etching step. In this case, the complete waveguide structure stays in full contact with the underlying substrate during the separation process. It is still fully supported and insensitive to the induced mechanical vibrations. After separating the microchips by standard sawing, the structures can be individually underetched as described.

In FIGs. 5.10 and 5.11, two scanning electron micrographs of microchips containing a single waveguide in the middle are shown. The micrographs were taken directly after the described fabrication processes.

Another problem involved with the fabrication process is the surface quality of the waveguide end facets (see FIG. 5.11a). The losses observed by coupling light in and out of the structures highly depend on this parameter. It is directly related to the surface roughness of these elements. Hence, special polishing techniques have to be applied to the waveguides. Due to the fragile nature of the suspended rib, a direct mechanical polishing of microchips is not possible. Instead, focused ion beam (FIB) processing was used to selectively cut the end facets of the waveguides. The cutting process is very precise and the resulting facets can become very smooth. The result of such a FIB polishing can be seen in FIG. 5.12.

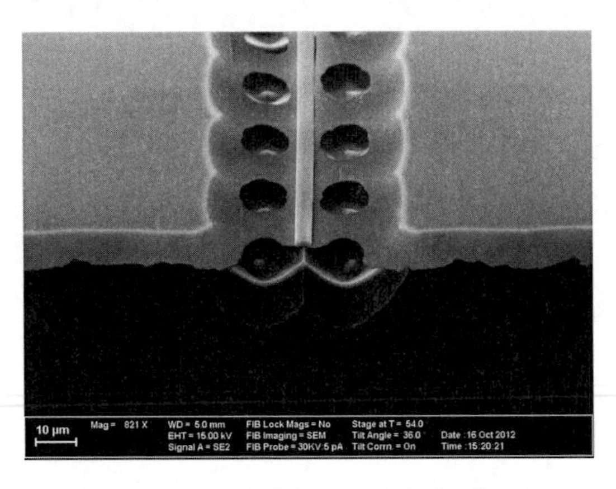

FIG. 5.12. Scanning electron micrograph of the presented silica-based waveguide design after cutting and polishing the end facets with a focused ion beam.

It is also possible to separate complete bars from the wafers. A bar consists of several microchips containing waveguides. The bars can be mechanically polished as single blocks before the last underetching process is performed. Afterwards, single microchips can be separated from the bars and individual etching processes can be applied. As such sequential methods are very time consuming, a more direct "polishing" method for waveguide facets is required. This can be carried out by using the same etching process which was already used to structure the sides of the waveguides. With the second lithographic mask, it is also possible to pre-define the end facets of this waveguides. In this case, the following etching process directly produces the required surfaces. A later dicing does not affect these areas. Thus, the sawing regions can be separated from the end facets and high quality surface finishes can be achieved. All the surfaces are chemically etched and thus a similar roughness level is produced at end facets and side walls of the waveguides. With this method, extensive post-production polishing can be avoided and manual processing steps are not required.

5.6 Summary and Outlook

Silica waveguides are of particular interest for a wide range of applications. However, the realization of suitable designs can be difficult depending on the specific material system. In systems where the refractive index n_w of the waveguide is smaller than the refractive index n_s of the substrate, no fully satisfying solution is actual available. The presented waveguide design combines simple manufacturing techniques with the exceptional optical properties of highly pure silica. It can further be used within various other material systems. The design allows the realization of elaborated optical circuits which can highly increase the functionalities of classical optical systems. The design can also be integrated with other waveguide structures in arbitrary material systems.

For example, homogeneous materials can be processed with the GOPHER method as described in the introductory section of this chapter. This technique was developed with silicon-based photonic crystal waveguides in mind, but it can also be applied to the presented waveguide design. The possibility of directly coupling short-range photonic crystals with long-range dielectric waveguides would highly increase the applicability of such elements and also the on-chip integration of photonic crystals would become possible. The two different waveguide designs are compatible to each other as in both cases the vertical confinement of the guided modes is due to total internal reflection in air. As the propagation constants in the photonic crystal are similar to the propagation constants in the dielectric regions, the mutual coupling of the two waveguides can be optimized by defining appropriate structural design parameters. With such technique, interconnections between distinct photonic crystals could be realized with intermediate dielectric waveguides. In FIG. 5.13, an example for such a hybrid design is illustrated. Furthermore, such a design can highly simplify the optical coupling to photonic crystal structures. Additional coupler elements, e.g., gratings or half-tapers, are not required when dielectric waveguides are used. These devices often work out of plane and allow mode coupling only at limited bandwidths. With a hybrid approach, well-known edge coupling techniques for conventional planar waveguide structures can be applied and also used for coupling to photonic crystal structures.

With silica and silicon, the presented waveguide design is based on a well-established technology and material platform. Optical microresonators with ultra-high Q factors can be produced. The practical applicability of such systems is widely demonstrated. Among other things, they are used for sensing applications, in quantum optics, and for spectroscopy. However, a satisfying solution for integrating optical microresonators with chip-based waveguide designs and couplers is still missing. The required coupling is often realized with conventional fiber tapers. These devices are not suitable for microintegration and other techniques have to be applied. By using the presented rib waveguide design for coupling, the problem can be solved and even complex integrated photonic circuits can be realized.

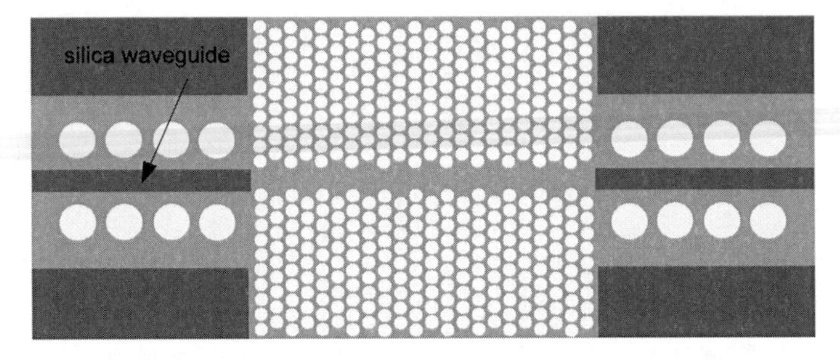

FIG. 5.13. Example of a combined system consisting of a photonic crystal waveguide and two silica waveguides. This structure can be realized in any heterogenic material system, but may also be implemented in fully homogeneous materials, e.g., in pure silicon by using the so-called GOPHER process [21].

For future applications, tapered waveguide structures are a basic requirement. Within such structures, the horizontal cross-section of the waveguide is modified along the direction of propagation. This allows a continuous change in the properties of the guided modes. Also a reversible transformation from multi- to single-mode operation is possible. For many applications in quantum optics, especially for the efficient coupling of microresonators and emitters to such waveguides, single-mode operation is preferred and small waveguide cross-sections are required. However, a small cross-section is highly challenging for efficient coupling. For optimization, the mode field diameters of the different elements have to be matched. By coupling to optical fibers, a multimode coupling region with a very large cross-section is required. In this case, the field overlap between the modes is maximized and the achievable coupling efficiency can be increased. For a transformation from multi- to single-mode operation, the waveguides must be tapered. The presented waveguide design allows such modifications and can be used to optimize the coupling conditions. Most typical are biconical tapered structures extending their cross-section at both ends of a waveguide. In FIG. 5.14, the energy distribution in such a device is exemplarily shown. The plot demonstrates that for the given specifications the influence of the taper length on the taper transmission can be neglected after a merely few micrometers of propagation.

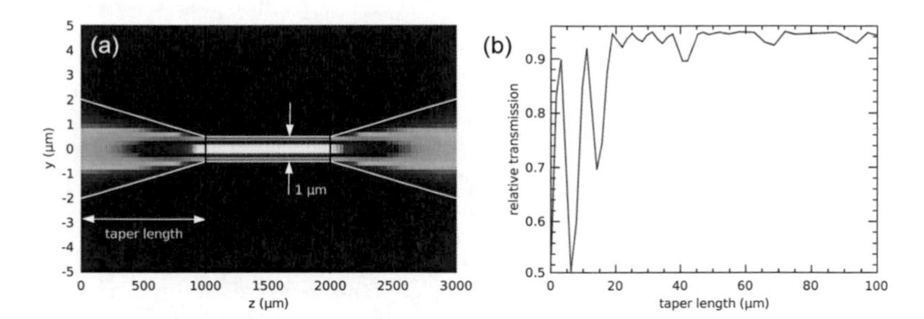

FIG. 5.14. (a) Simulation of the energy distribution of the fundamental TM mode in a biconical tapered waveguide. The three sections are each 1000 µm long. The waveguide has a height of 2 µm and is tapered down from 4 µm to 1 µm. The simulation wavelength was 780 nm. In this configuration, 95 % of the initial energy is coupled into the fundamental mode of the output port. (b) Taper transmission vs. taper length.

The discussed tapered structures are also required for the realization of other basic photonic elements. Due to the decrease of the waveguide cross-section, the evanescent part of the guided fields, which is located on the outside of the waveguides, can be extended. This is especially relevant for evanescently coupling light to other structures or elements. An example is coupling to silica-based integrated microresonator systems; another example is the realization of beam splitters and combiners. In integrated photonics, directional couplers are widely used. In this type of couplers, two optical waveguides are located nearby and parallel to each other. Due to the interaction of evanescent fields in both waveguides, intense cross-coupling between these two distinct elements can be achieved. Depending on the length of the coupler section and the strength of their evanescent interaction, the guided fields can be swept in a highly controlled manner. These couplers can be fabricated to allow arbitrary splitting ratios between the two distinct waveguides.

This coupler design allows even more advanced photonic elements to be realized, e.g., implementing of a fully integrated Mach-Zehnder interferometer on chip-level. In these devices, directional couplers are arranged in tandem such that their connecting waveguides include a fixed path difference. By recombining the two paths, interference takes place and phase sensitive applications can be realized. Furthermore, active path difference control can be included by thermally tuning one arm of the interferometer with additionally integrated heating elements. In this case, optical modulators or active switches can be implemented.

In the field of quantum optics, the presented waveguide design can be used for the realization of a complete on-chip version of a Hanbury Brown and Twiss setup (see Appendix C). After out-coupling single photons from the two output ports of an integrated beam splitter, photon detection can be carried out by directly connected avalanche photo diodes (APDs). In this way, complete on-chip implementations of quantum optics experiments can be realized. Furthermore, simple lab-on-chip designs could be extended to highly complex optical quantum networks by cross-connecting several of such devices.

For some classical applications, such designs are already state of the art. They are foremost used within the telecommunication industry and allow connecting several hundreds of individual optical components to each other. For such applications, the structures are mostly fabricated on SOI substrates. Therefore, the available wavelength range is limited to the near infrared spectral range. With the presented rib waveguide design, the functionalities of PICs could be transferred to other material systems and wavelength ranges. The main advantage of thermally oxidized silica is the high purity of the material. With the presented waveguide design, the applications for PICs could be easily extended into the visible spectral range. This design approach is also perfectly suited for applications working in the ultraviolet spectral range.

[1] M. Ménard and A. Kirk, "Broadband integrated Fabry-Perot electro-optic switch," in *International Conference on Photonics in Switching (PS)*, 2008, pp. 1–2.

[2] N. Sherwood-Droz, H. Wang, L. Chen, B. G. Lee, A. Biberman, K. Bergman, and M. Lipson, "Optical 4x4 hitless silicon router for optical networks-on-chip (NoC)," *Opt. Express*, vol. 16, no. 20, pp. 15915–15922, 2008.

[3] D.-X. Xu, M. Vachon, A. Densmore, R. Ma, A. Delâge, S. Janz, J. Lapointe, Y. Li, G. Lopinski, D. Zhang, Y. Q. Liu, P. Cheben, and J. H. Schmid, "Label-free biosensor array based on silicon-on-insulator ring resonators addressed using a WDM approach," *Opt. Lett.*, vol. 35, no. 16, pp. 2771–2773, 2010.

[4] J. Bauters, M. Heck, D. John, D. Dai, M. Tien, J. S. Barton, A. Leinse, R. Heideman, D. J. Blumenthal, and J. E. Bowers, "Ultra-low-loss high-aspect-ratio Si_3N_4 waveguides," *Opt. Express*, vol. 19, no. 4, pp. 3163–3174, 2011.

[5] A. Gorin, A. Jaouad, E. Grondin, V. Aimez, and P. Charette, "Fabrication of silicon nitride waveguides for visible-light using PECVD: a study of the effect of plasma frequency on optical properties," *Opt. Express*, vol. 16, no. 18, pp. 13509–13516, 2008.

[6] W. Bogaerts, D. Taillaert, and B. Luyssaert, "Basic structures for photonic integrated circuits in silicon-on-insulator," *Opt. Express*, vol. 12, no. 8, pp. 1583–1591, 2004.

[7] B. D. Timotijevic, F. Y. Gardes, W. R. Headley, G. T. Reed, M. J. Paniccia, O. Cohen, D. Hak, and G. Z. Masanovic, "Multi-stage racetrack resonator filters in silicon-on-insulator," *J. Opt. A Pure Appl. Opt.*, vol. 8, no. 7, pp. 473–476, 2006.

[8] K. Rivoire, A. Faraon, and J. Vuckovic, "Gallium phosphide photonic crystal nanocavities in the visible," *Appl. Phys. Lett.*, vol. 93, no. 6, pp. 063103–063103, 2008.

[9] A. Majkic, M. Koechlin, G. Poberaj, and P. Günter, "Optical microring resonators in fluorineimplanted lithium niobate," *Opt. Express*, vol. 16, no. 12, pp. 8769–8779, 2008.

[10] T. Stievater, W. Rabinovich, D. Park, J. B. Khurgin, S. Kanakaraju, and C. J. K. Richardson, "Low-loss suspended quantum well waveguides," *Opt. Express*, vol. 16, no. 4, pp. 2621–2627, 2008.

[11] Y. Hanada, K. Sugioka, and K. Midorikawa, "UV waveguides light fabricated in fluoropolymer CYTOP by femtosecond laser direct writing," *Opt. Express*, vol. 18, no. 2, pp. 446–450, 2010.

[12] A. Himeno, K. Kato, and T. Miya, "Silica-based planar lightwave circuits," *J. Sel. Top. Quantum Electron.*, vol. 4, no. 6, pp. 913–924, 1998.

[13] T. Miya, "Silica-based planar lightwave circuits: Passive and thermally active devices," *J. Sel. Top. Quantum Electron.*, vol. 6, no. 1, pp. 38–45, 2000.

[14] R. Henze, A. Thies, and O. Benson, "Wellenleiteranordnung," DE 10 2012 222 898.5, 2012.

[15] Z. Sheng, B. Yang, L. Yang, J. Hu, D. Dai, and S. He, "Experimental Demonstration of Deeply-Etched SiO_2 Ridge Optical Waveguides and Devices," *J. Quantum Electron.*, vol. 26, no. 1, pp. 28–34, 2010.

[16] D. Dai and Y. Shi, "Deeply Etched SiO_2 Ridge Waveguide for Sharp Bends," *J. Light. Technol.*, vol. 24, no. 12, pp. 5019–5024, 2006.

[17] M. Popovic, K. Wada, S. Akiyama, H. A. Haus, and J. Michel, "Air trenches for sharp silica waveguide bends," *J. Light. Technol.*, vol. 20, no. 9, pp. 1762–1772, 2002.

[18] M. Duguay, Y. Kokubun, T. Koch, and L. Pfeiffer, "Antiresonant reflecting optical waveguides in SiO_2-Si multilayer structures," *Appl. Phys. Lett.*, vol. 49, no. 1, pp. 13–15, 1986.

[19] J.-P. Laine, B. E. Little, D. R. Lim, H. C. Tapalian, L. C. Kimerling, and H. A. Haus, "Microsphere resonator mode characterization by pedestal anti-resonant reflecting waveguide coupler," *Photonics Technol. Lett.*, vol. 12, no. 8, pp. 1004–1006, 2000.

[20] Y. Sugimoto, Y. Tanaka, N. Ikeda, Y. Nakamura, K. Asakawa, and K. Inoue, "Low propagation loss of 0.76 dB/mm in GaAs-based single-line-defect two-dimensional photonic crystal slab waveguides up to 1 cm in length," *Opt. Express*, vol. 12, no. 6, pp. 1090–1096, 2004.

[21] S. Hadzialic, S. Kim, A. S. Sudbo, and O. Solgaard, "Two-Dimensional Photonic Crystals Fabricated in Monolithic Single-Crystal Silicon," *Photonics Technol. Lett.*, vol. 22, no. 2, pp. 67–69, 2010.

[22] T. Saito, T. Hanada, N. Kitamura, and M. Kitamura, "Photosensitivity in silica-based waveguides deposited by atmospheric pressure chemical vapor deposition," *Appl. Opt.*, vol. 37, no. 12, pp. 2242–2244, 1998.

[23] J. Hu and C. Menyuk, "Understanding leaky modes: slab waveguide revisited," *Adv. Opt. Photonics*, vol. 1, no. 1, pp. 58–106, 2009.

[24] A. A. Bettiol, S. V. Rao, E. J. Teo, J. A. van Kan, and F. Watt, "Fabrication of buried channel waveguides in photosensitive glass using proton beam writing," *Appl. Phys. Lett.*, vol. 88, no. 17, p. 171106, 2006.

[25] K. Chiang, "Dispersion characteristics of strip dielectric waveguides," *Trans. Microw. Theory Tech.*, vol. 39, no. 2, pp. 349–352, 1991.

[26] M. de Laurentis, A. Irace, and G. Breglio, "Determination of single mode condition in dielectric rib waveguide with large cross section by finite element analysis," *J. Comput. Electron.*, vol. 6, no. 1–3, pp. 285–287, 2007.

[27] A. Melloni, F. Carniel, R. Costa, and M. Martinelli, "Determination of bend mode characteristics in dielectric waveguides," *J. Light. Technol.*, vol. 19, no. 4, pp. 571–577, 2001.

[28] C. Pollock and M. Lipson, *Integrated Photonics*. Norwell: Kluver, 2003.

[29] E. Marcatili, "Dielectric rectangular waveguide and directional coupler for integrated optics," *Bell Syst. Tech. J*, vol. 48, no. 7, pp. 2071–2102, 1969.

[30] A. Kumar, K. Thyagarajan, and A. Ghatak, "Analysis of rectangular-core dielectric waveguides: an accurate perturbation approach," *Opt. Lett.*, vol. 8, no. 1, pp. 63–65, 1983.

[31] J.-S. Lee and S.-Y. Shin, "On the validity of the effective-index method for rectangular dielectric waveguides," *J. Light. Technol.*, vol. 11, no. 8, pp. 1320–1324, 1993.

[32] K. S. Chiang, "Analysis of rectangular dielectric waveguides: effective-index method with built-in perturbation correction," *Electron. Lett.*, vol. 28, no. 4, pp. 388–390, 1992.

[33] I. Malitson, "Interspecimen comparison of the refractive index of fused silica," *J. Opt. Soc. Am. B*, vol. 55, no. 10, pp. 1205–1208, 1965.

[34] J. Ruano, V. Benoit, J. S. Aitchison, and J. M. Cooper, "Flame hydrolysis deposition of glass on silicon for the integration of optical and microfluidic devices," *Anal. Chem.*, vol. 72, no. 5, pp. 1093–1097, 2000.

Chapter 6

Polymer-Based Whispering Gallery Mode Resonators and Waveguides

6.1 Introduction

In modern photonics, waveguides are the key component for the development of novel integrated devices. As substrate materials, polymers are of increasing interest [1]. Their physical properties can often be changed by simple light exposure. By modifying their specific composition, they can be tuned for a wide range of applications, e.g., for using as thermo-optic devices [2], as electro-optic devices [3], and as optical light emitting diodes [4]. Polymers can be molded in various forms or may be designed to remain flexible after preparation [5]. Hence, they are the ideal basis for ultra-cheap photonics. They are also the material of choice for building chip-based interconnects to photonic circuit boards (PCBs) [6]. Polymers are thermally and mechanically robust, even the integration with classical semi-conductor components has been demonstrated [7].

For structuring such materials, three dimensional (3D) direct laser writing (DLW) can often be applied. The time and cost saving combination of the preparation and manufacturing processes is an important benefit compared to conventional wafer-based etching and deposition techniques. DLW schemes offer a substantial reduction in the total production costs [8]. Furthermore, most of the DLW techniques do not demand on a high-class clean room environment or other expensive lithography equipment. Waveguides can be produced deeply buried inside the materials, thus the deposition of additional capping layers is not required. For direct waveguide writing, primarily non-linear excitation by infrared two-photon absorption is utilized [9]. With such a technique, even the self-confined fabrication of self-written waveguides could be demonstrated [10]. However, two-photon absorption typically uses a time consuming voxel-based writing scheme and requires expensive pulsed laser sources.

A less demanding continuous direct writing technique is based on single photon absorption in so-called diffusion-mediated photopolymers [11]. In such materials, the refractive index is locally changed by initiating intensity-dependent polymerization processes. Due to the higher molecular weight of the produced polymers, an increase of the refractive index is observed. The index contrast is further enhanced by diffusion processes which only affect the distribution of lighter monomers. By this effect, very high polymer densities are resulting and well localized refractive index profiles can be written. Within such materials, the self-writing of straight waveguide structures has also been demonstrated [12].

In the following chapter, these two different types of laser writing are applied to directly structure waveguides and other optical components into polymers. These are namely the two-photon absorption based DLW lithography and the DLW lithography in diffusion-mediated photopolymers. Both presented writing schemes provide very general technological approaches to the realization of integrated optical components for applications in integrated quantum photonics and other fields of optics. Within this chapter, two application examples are presented. The realization of an integrated photonic circuit based on a functionalized photopolymer containing active quantum emitters is demonstrated. Furthermore, the direct fabrication of polymer waveguides with a novel DLW scheme shows the applicability of such polymer designs for the realization of chip-based interconnections. The chapter starts with an introduction to important material properties of polymers and summarizes the processing techniques applicable to such materials.

6.2 Properties of Optical Polymers

Optical polymers possess many features which distinguish them from conventional material systems and make them highly interesting for optical applications. Beside simple fabrication techniques and the relatively low price of polymers, foremost their versatile applicability and a variety of different processing schemes are outstanding features of polymer-based photonics. Polymers can be specifically modified in their physical and optical properties and possible applications can span a wide range from simple low cost devices to elaborated photonic elements. In the following section, the most relevant properties of optical polymers are summarized. The full physical and chemical background of such materials can be found in [5].

6.2.1 Refractive Index

A main advantage of optical polymers is the wide range of refractive indices observable within such organic materials. This allows choosing the right composition for specific applications and matching polymer-based systems to other non-polymeric optical components. Depending on the operational wavelength, typical values for the refractive index of polymers are in the range between 1.3 and 1.7 [13]. It is also possible to combine compatible polymers to achieve a targeted fine-tuning of the refractive index in the resulting mixtures.

In polymers, essentially three material parameters are responsible for the refractive index [14]. These are the free material volume, the electric polarizability of the functional groups and the band gap between the optical wavelength and absorption bands inside the material. Thus, a systematic control of the refractive index in polymers can be achieved by structural modifications of the contained molecules, by varying their electric properties, or by adding special guest polymers. Due to a decrease in the free material volume, the molecular packing density is increased and the refractive index can be shifted to higher values. High temperature densification can also be used to raise the refractive index of the material [15]. Furthermore, by adding fluorine or deuterium atoms to the polymer molecules, the absorption bands can be substantially shifted. This can be used for broadening the transparent wavelength window or to lower the refractive index in the vicinity of standard absorption bands.

Birefringence

In polymers, birefringence can be controlled by selecting suitable polymer chains. The possibility of molecular engineering is a significant advantage over conventional material systems where it is not possible to considerably change the anisotropy. With three-dimensional cross-linked polymers, extremely low values of birefringence can be achieved. In such materials, nearly no molecular orientation takes place during all kinds of processing [16]. On the other hand, with some specific aromatic polymers, very high values of birefringence are also possible to achieve (up to 0.24) [17]. They are mainly caused by naturally occurring molecular orientations along specific spatial axes of the material.

Temperature Dependence

The refractive index of polymer materials normally shows very strong temperature dependence. The corresponding thermo-optic coefficient $\beta = \mathrm{d}n/\mathrm{d}T$ is commonly around $-10^{-4}\ \mathrm{K}^{-1}$ and thus nearly one order of magnitude larger as for standard inorganic glasses [18]. Furthermore, such polymer materials often show negligible thermal conductivity. This makes them perfectly suited for simple and energy efficient realizations of thermo-optic switching devices [19]. By adding special heating elements inside or on top of the polymer, the material can be heated at certain spots and the thermal expansion of this structure can be used for tuning. With this technique, coupling between two distinct waveguides can be controlled by simply changing their spacing via thermal control.

An important difference of polymers compared to conventional optical materials is the negativity of their thermo-optic coefficients [19]. Heating such materials causes a decrease in the refractive index due to the increase of the free material volume. This allows designing athermal optical components where the optical propagation properties of waveguides or resonators are not influenced by external temperature changes. By combining conventional silicon-based optical core structures with a suitable polymer as cladding material, the positive thermo-optic coefficient of silicon can be canceled out by the negative thermo-optic coefficient of the used polymer. By that, the propagation constants of guided waves become independent on temperature and athermal behavior of waveguide transmission is achieved.

Wavelength Dependence

For most waveguide applications, broadband transmission properties are required. In these cases, the wavelength dependence of the material should be as low as possible. For most polymers designed for optical applications, the material dispersion $\mathrm{d}n/\mathrm{d}\lambda$ is in the range of 10^{-6} nm^{-1} [20]. This is comparable to the dispersion in silica-based systems but much lower as for doped glasses or other semiconductor materials.

6.2.2 Optical Loss

By using polymer-based photonic elements in practical applications, optical losses are most relevant. In general, they are required as low as possible even though their values can not compete with crystalline materials or glasses. Depending on the intended wavelength range, individual material components can be changed to achieve a transmission maximum in the working range of the fabricated structures. The three major sources of losses in polymer optical waveguides are material absorption, scattering loss, and fiber coupling loss [21].

Absorption loss

Most of the available polymers for optical applications show a relatively high material absorption in the ultraviolet (UV) spectral range. This is predominantly due to electronic excitations of molecules inside the polymer. In the visible (VIS) or near infrared (NIR) spectral ranges, most polymers are transparent and merely weak absorption by vibrational or single-triplet excitation processes is observed. Direct electronic excitation, like in the UV spectral range, is normally not an issue within these ranges. Hence, it is possible to reduce the absorption losses by simply modifying the polymeric bounds. This changes the vibrational spectrum of the molecules and shifts the vibrational excitation losses out of the desired transmission window. With this technique, minimum absorption losses around 0.1 dBcm^{-1} could be achieved in the VIS and NIR spectral ranges [13].

Scattering Loss

The scattering losses of polymers are mainly caused by different defect centers located inside the material. These centers can be foreign particles, material voids, cracks, or bubbles. As these defects are commonly much larger in size than the used optical wavelength (>> 1 μm), the process is mainly based on geometric wavelength independent scattering. If the dimensions of a defect center are in the range of the transmitted light (~ 1 μm), the main scattering effect is based on Mie scattering and becomes slightly wavelength dependent. Examples for such kind of defects are foreign particles, dust, dissolved bubbles, and unreacted monomers. Other sources of scattering loss are density fluctuations and inhomogeneities in the material composition. These defects are dominant on length scales way below the wavelength of the transmitted light (<< 1 μm) and result in highly wavelength dependent losses due to Rayleigh scattering. As all three effects have different wavelength characteristics, the different sources of scattering can be deduced by measuring the transmission losses at different wavelengths. The resulting data can then be fitted to an empirical law given by [5]

$$\alpha_{scattering} = A + B/\lambda^2 + D/\lambda^4 \tag{6.1}$$

with A as the coefficient describing the losses caused by large particles (geometrical scattering), B as the loss coefficient for Mie scattering, and D as the Rayleigh scattering coefficient. From this equation, the main source of loss can be identified and possibly reduced.

In addition to these intrinsic scattering losses, further loss mechanisms can occur from the accuracy of the structuring process used to produce waveguides or resonators. The main factor of such transmission dependent losses is the roughness of the written structures. Their quality can be increased by tempering the polymers after processing. With actual polymer structuring techniques, minimum propagation losses of around 0.03 $dBcm^{-1}$ can be achieved [13].

A way to reduce the scattering losses in polymer materials is by ensuring high purity and homogeneity of the used materials. It is important to follow ultra-clean processing procedures starting from the preparation of polymer compositions up to final structuring and fixation steps. Another important factor is the stress-free applicability of polymer materials onto arbitrary substrates. Stress-induced scattering losses may occur from high refractive index differences established inside the polymer. This phenomenon can also be avoided by carefully tempering the material after preparation.

Fiber Pigtail Loss

For coupling polymer-based waveguide designs to glass fibers or other external optical components, the different propagation modes of the two systems have to be matched. The geometric size of the waveguides must correspond to the size of fiber modes. The refractive index of the polymer material and the refractive index contrast between waveguide cores and possible claddings must also be considered. Furthermore, for efficiently coupling different optical components, a precise and robust alignment has to be ensured. By using an index-matching gel or by adding antireflection coatings onto the waveguide facets, coupling losses due to reflections can be minimized. Also the mechanical surface quality of the waveguide facets is a very important parameter for achieving high coupling efficiencies. For avoiding extensive back-scattering, at least a surface roughness of $\lambda/10$ is required.

6.2.3 Environmental Stability

Polymers are complex chemical compounds based on organic molecules. Under certain environmental conditions, these materials may undergo some specific aging effects not observable in other conventional material systems used for optical applications. However, by modifying the chemical structure of the contained monomers and with controlling the linkage between the different material components, the physical properties of the resulting polymers can be widely changed. This controllability includes parameters like hardness, toughness, water uptake, or stability against aging. Other important properties to take into account are the temperature dependence of the material and the effects of high-energy radiation like UV light exposure over extended periods of time. For the telecommunication industry, the environmental stability is often defined by the so-called Telcordia requirements [22]. These reliability tests for optoelectronic components can directly be transferred to the stability requirements of chip-integrated photonic devices based on polymers.

6.3 Processing

For processing polymer-based material systems, a number of different methods were developed. In contrast to conventional waveguide materials, most optical polymers can be processed as liquids or solids. This highly extends the possibilities of applying such materials to different substrates. The available structuring methods are also highly enhanced compared to other material systems. A major advantage of most polymers is their high elasticity. This allows application and structuring of such materials even on flexible substrates, e.g., Mylar films. An application example are flexible polarization transformers used in holographic and coherent optical systems [23].

The structuring methods for polymers can be distinguished in two different approaches. They can be performed by various lithographic methods out of a thin film or within a thick sample, or by direct structuring via molding or printing. A special type of the latter is so-called soft lithography. Thereby, the structuring of the polymer is performed during the application process by directly molding the polymer in the appropriate form which is produced with a polydimethylsiloxane (PDMS) stamp. It already contains the intended microstructure and can be used for micromolding in capillaries [24] or microcontact printing [25]. The stamp itself is produced by casting from a silicon master which is patterned with standard lithography methods. Soft lithography allows simple and cost efficient mass production of optical waveguides as the structuring process can be repeated without any tool wearing.

By using polymer materials for optical circuitry, the homogeneous and controlled application onto various substrates is a particular challenge. For all lithographic structuring processes, uniform material thicknesses and bubble- and grove-free application techniques are required. Furthermore, the applied polymers have to show very good adhesion to the substrates, but must still be applicable in good optical quality. Another important aspect is the achievable material thickness for the polymer films. While with molding processes nearly arbitrary thicknesses can be produced, conventional application methods are limited to relatively thin material thicknesses.

Methods for applying polymer films onto nearly arbitrary substrates are spin coating, extrusion, lamination, and doctor blading. Each of these methods has advantages and disadvantages concerning the achievable film quality, processing speed, maximum film thickness, and application homogeneity. The by far most common application technique is spin coating. This easy and reliable method produces polymer films of excellent quality with material thicknesses of up to several hundred micrometers. It is a well-known technique and can be used for a variety of optical polymers.

For the fabrication of waveguide structures in polymer films, different techniques are available. In hot embossing [26] or with so-called nanoimprinting techniques [27], a pre-structured stamp with the desired optical microstructure is pressed into the polymer films. Another example is laser ablation where the structures are produced by evaporating excess material with a focused laser beam [28].

It is also possible to use structuring techniques based on well-known semiconductor processing [8]. A photoresist can be used as structuring mask by applying a thin film directly on top of the optical polymer. Afterwards, the photoresist can be patterned by various structuring methods (e-beam, mask-patterning, imprint techniques). After developing the patterned resist, the structures are washed off by a solvent. The structuring of the polymer film can then be carried out by a variety of etching methods known from semiconductor industry. Therefore, reactive ion etching or plasma etching are often applied [29].

Another method for processing polymer films is using some sort of direct lithography. For instance, in dye-doped optical polymers, structuring can be performed by thermal UV bleaching [30]. Also direct structuring of waveguides via field poling has been successfully demonstrated [31]. Furthermore, it is also possible to directly pattern polymers via standard e-beam lithography [32].

6.3.1 Direct Laser Writing (DLW)

Photosensitive systems are defining a special class of materials in optical polymer processing. They are often simply referred to as photopolymers. Within such materials, the polymerization process of contained monomers is initiated by means of precisely directed light exposure. Photopolymers can be prepared and processed with large material thicknesses allowing volume bulk processing. DLW is a maskless structuring method. It not only allows the creation of simple planar waveguide structures but also the realization of highly complex three-dimensional (3D) optical elements. By being relieved from time consuming and expensive masked-based processing schemes, fast and cost efficient structuring processes are possible also for extended polymer-based 3D photonic designs. For example, DLW techniques allow the direct fabrication of continuous waveguide structures several meters in length not possible with other common techniques. They can be written on large substrates, e.g., flexible rolls of plastic, produced by polymer roll-to-roll processing [28].

In nearly all photopolymers, the core region is created by an increase of the local refractive index. This is carried out by initiating localized polymerization processes of monomers and oligomers as the main constituents of undeveloped materials. Another basic component are photosensitive initiator molecules. They are optically activated and responsible for starting the polymerization process. The molecules produce acids which act as chemical linkage for monomers during polymerization. Due to the limited availability of generated acids and contained monomers, the polymerization process is mainly limited to the exposed areas. Hence, well defined refractive index structures can be generated inside the material. In principal, two different writing schemes can be distinguished for creating waveguides in photopolymers. The methods mainly differ in the required laser power to start the polymerization process and how the refractive index gradient of the guiding structures is established.

DLW via Two-Photon Absorption

The first direct laser writing scheme bases on two-photon absorption in photopolymers. By focusing intense femtosecond laser pulses into such photopolymerizable materials, energy is absorbed by contained photoinitiator molecules which can only be activated when two photons out of the laser field are acquired. Single-photon absorption is not sufficient to initiate the polymerization process. The two-photon absorption can be performed simultaneously or by using some intermediate electronic levels of the molecules (see FIG. 6.1). For standard DLW, only simultaneous absorption is used. The effective cross-sections of such two-photon absorption processes are very small and thus the required high energy densities can only be achieved in the focus of a laser beam. Therefore, the initiated polymerization process is highly localized to the near and direct vicinity of the focus. The method allows high spatial resolution in 3D because the corresponding trigger level can be controlled by simply changing the laser power.

Single dots, lines, planes, or even complete 3D structures can be written voxel by voxel with sequentially scanning the focal spot through the full material volume. After completely polymerizing the desired structures, the other material portions still contain large amounts of residual monomers. They are normally washed off and only the written structures remain.

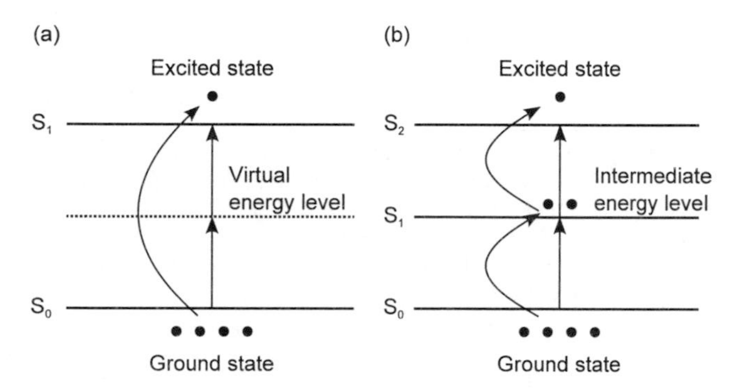

FIG. 6.1. Illustration of two-photon absorption (TPA) (from [9]). (a) Simultaneous TPA for direct excitation. (b) Stepwise TPA over an intermediate electronic energy level.

The described method allows simple fabrication of planar waveguide structures on arbitrary substrates, but it is also perfectly suited for producing complex 3D structured optical elements. With this technique, the fabrication of complex photonic crystal structures could be demonstrated [33]. It is also possible to produce arbitrary shaped non-optical microstructures like single micro-gearwheels or further micromechanical components. Due to the absorption properties of two-photon activated photopolymers, the polymerization process is commonly started by using near infrared pulsed femtosecond lasers at wavelengths around 800 nm. By choosing proper excitation parameters for spot size, energy density, and laser power, typical lateral feature sizes are around 80 nm [34]. However, the achievable resolution as well as the minimum feature size can be further enhanced by using other materials and techniques. By using stimulated emission depletion microscopy (STED), the possible feature sizes can be reduced to less than 40 nm [35]. In this case, the excitation scheme is modified by changing the contained photoinitiator to support STED.

DLW in Diffusion-Mediated Photopolymers

Another possibility of direct laser writing is given in so-called diffusion-mediated photopolymers. Also in such materials the localized polymerization process is started by selectively exposing the material to appropriate laser light. But in contrast to the previous type of photopolymers, these materials are optimized to produce an intrinsic refractive index gradient solely by diffusion without the requirement for washing off unexposed polymer portions.

For diffusion-mediated photopolymers, a high mobility level of contained monomers is characteristic. In such materials, the photoinitiator can often be activated by continuous wave laser excitation. As the process is triggered by single-photon absorption, only moderate laser powers are required. The contained photoinitiators are often sensitive within the UV spectral range. By absorbing photons, spatially localized polymerization processes are started. In diffusion-mediated materials, the mobility of polymers is much lower as compared to the mobility of the smaller monomers. Thus, a concentration gradient is established between the monomers within illuminated and non-illuminated volumes. Directly created polymers are nearly fixed at their specific material positions. Monomers inside the neighboring regions start to diffuse into the illuminated areas and get also polymerized. Due to this effect, more and more polymer molecules are created inside the illuminated volumes and the local polymer density highly increases. Due to this effect, the free material volume is also locally increased and the corresponding refractive index in this region rises. When a sufficiently large refractive index contrast between illuminated and non-illuminated volumes is achieved, residual photoinitiators can be bleached out in a bulk illumination process with UV light. During this process, all residual monomers are completely polymerized without further diffusion and a stable cross-linked polymer matrix is generated. This process also slightly increases the overall refractive index of the material, but keeps the previously established refractive index contrast between illuminated and non-illuminated volumes basically unaffected. After bulk illumination, the material is completely developed and insensitive against any further light exposure.

In diffusion-mediated photopolymers, maximum refractive index contrasts of up to $5 \cdot 10^{-3}$ can be achieved in the VIS and NIR spectral ranges. Thereby, the final refractive index profile depends on illumination time t and writing intensity I. An applicable empirical relation is given by $\Delta n \approx I^{\alpha} t$, with the exponent α often between one half and one [36]. In diffusion-mediated photopolymers, the refractive index contrast is limited by material properties. Thus, only weak guiding conditions comparable to standard glass-fibers are achievable. The low index contrast limits the possible bending radii of curved waveguide structures, but highly simplifies direct coupling between fibers and polymer-based optical elements.

In addition to the ability of structuring simple planar optical waveguides on arbitrary substrates, this writing technique also allows the limited fabrication of more complex 3D optical elements. As the writing process is carried out by simple continuous wave laser excitation, this lithography method is particular interesting for producing large continuous structures in high speed. With such a structuring method, writing velocities of up to 20 m/s were demonstrated [36]. Furthermore, by simply changing the shape or intensity of the writing beam, extended tapered waveguides or other continuously varying structures can be produced. Another asset of such diffusion-mediated polymer materials is the possibility to directly incorporate optical gratings or holographic structures inside the waveguide designs. In this way, rather complex integrated optical systems containing wavelength dependent filtering or beam shaping can be produced.

The major advantage of direct laser writing in diffusion-mediated photopolymers is that the created waveguide structures are inherently buried inside a bulk material. However, in many diffusion-mediated photopolymers also a different solubility of structured and unstructured material portions can be observed. In such materials, the higher polymer density inside the written structures allows an exposure by selectively dissolving all surrounding regions with lower polymer density. Hence, also with this DLW technique the creation of free standing polymer-based optical elements is possible by simply washing off unexposed material portions after finishing the writing processes.

6.4 Direct Laser Writing of Functionalized Photonic Elements

A major advantage of polymers as basic material in photonics is their simple miscibility with further organic or inorganic components. This allows a precise controllability and targeted modification of nearly all relevant material properties. In conventional material systems, complex and burdensome growth or implantation techniques are required for the heterogenic integration of extraneous materials or a controlled doping. In contrast, for polymers a homogeneous distribution of various other components can often be realized by simple mixing and stirring processes. If the miscibility between the polymer and additional components is sufficiently large, simplest material integration can be realized. It is also possible to direct functionalize polymers by this method. Thus, for specific applications, highly optimized materials can be developed. An example of such hybrid materials are so-called polymer-dispersed liquid crystals (PDLCs) [37]. In such materials, integrated electro-optic switches can be realized by electrostatic poling. Another example for functionalizing polymers is the integration of dye molecules inside the material [38]. This even allows the realization of polymer-based amplifiers and frequency converters for applications in active photonics.

Another possibility for functionalizing polymers is the direct integration of active quantum emitters. With such materials, the applicability of polymer-based photonic elements can also be extended to fields of experimental or applied quantum optics. In the following section, such a functionalized material is presented and two basic elements for future integrated quantum photonics, the so-called nanodiamond-dispersed arc waveguides and disk resonators, are demonstrated. As the applied technology is based on a two-photon writing process, the high spatial resolution of the method allows precise control over the structured elements. For a first characterization of the material, the optical properties of an integrated resonator system were measured.

The presented method of functionalizing polymers for applications in quantum optics was developed in cooperation with the Karlsruhe Institute of Technology (KIT). The analysis of the optical properties of the hybrid system and the demonstration of first optical elements resulted in a publication [39].

6.4.1 Writing Method

The photonic structures are produced from a material which is based on the monomer pentaerythritol tetraacrylate (PETA). As photoinitiator system, the organic molecule 7-diethylamino-3-thenoylcoumarin is used. The corresponding polymerization inhibitor is monomethyl ether hydroquinone. Furthermore, for functionalizing the material, an ethanol-based nanodiamond suspension was added to the mixture. These nanodiamonds have a medium diameter of 25 nm and include nitrogen vacancy (N-V) defect color centers as quantum emitters (see also Chapter 4.5.1).

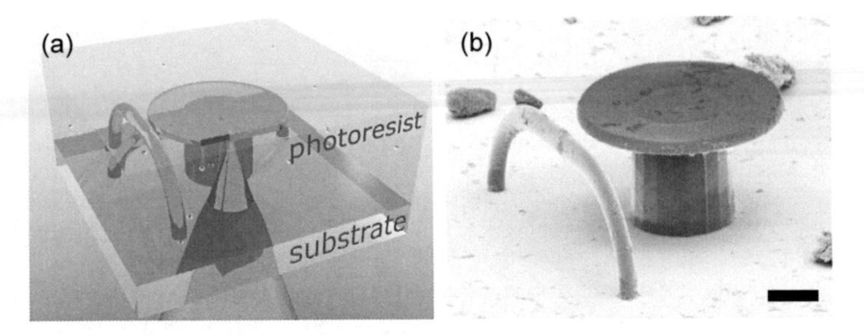

FIG. 6.2. Direct laser writing process (from [39]). (a) A femtosecond laser beam is focused into the photoresist and polymerizes monomers within the focus via two-photon absorption. This allows voxel by voxel writing of highly complex 3D structures. (b) Scanning electron micrograph of the illustrated structure. An arc waveguide is shown next to the disk resonator. The scale bar is 5 µm.

The material is structured by a pulsed Ti:sapphire laser emitting at a wavelength of 810 nm. The laser pulses have a duration of 100 fs. During the writing process, an average power of around 6 mW is deposited. For a tight focusing of the laser beam, an oil immersion lens (100x, NA 1.40) is used. During the writing process, the position of the sample is controlled in 3D by a piezoelectric nanopositioning system. With this method, structuring velocities of up to 50 µm/s are achievable.

The laser pulses are directed into the material. A localized polymerization process of monomers is thereby initiated by activating the photoinitiators via two-photon absorption (see also Section 6.3.1). During the writing process, the solubility of the material is changed. The polymerized structures become insoluble to isopropyl alcohol. Thus, permanent structures can be written by simply washing off the undeveloped areas after the writing processes are finished. The residual monomers are dissolved and only the polymerized structures remain. In FIG. 6.2a, the schematic of the writing process is shown. A scanning electron micrograph of the resulting polymer microstructures is presented in FIG. 6.2b.

6.4.2 Photonic Elements

Photonic elements are exposed to the environment after finishing the writing processes. Hence, the final structures are uncladded and exhibit a high refractive index contrast. In the visible spectral range, a ratio of 1:1.5 is typical. Due to this high contrast, small bending radii can be realized for the written elements without introducing large amounts of radiative loss. This allows highly compact designs for integrated waveguide and resonator systems with smallest structural footprints. The high refractive index contrast of the produced polymer-based photonic elements is analog to the achievable index contrast in air-cladded silica resonator systems.

Microresonators

For a simple proof-of-technology demonstration, diamond-dispensed polymer disk-type microresonators with diameters of 20 µm and silica layer thicknesses of around 1.2 µm were produced by the described DLW method. The stems underneath the disks had overall diameters of 10 µm and heights of 15 µm. Hence, sufficient separation between the circulating modes, and the substrate or stem is ensured. In FIG. 6.2b, a scanning electron micrograph of such a diamond-dispersed polymer microresonator is shown in conjunction with some 3D arc waveguides.

By assuming a polymer refractive index of $n = 1.5$ and a resonance wavelength of $\lambda = 770$ nm, numerical simulations with a commercial mode solver (Photon Design Fimmwave) provide for the fundamental TM mode an effective refractive index of $n_{eff}^{TM} = 1.4468$. For the fundamental TE mode $n_{eff}^{TE} = 1.4427$ was found. In FIG. 6.3, the corresponding intensity mode profile is shown for the fundamental TM mode.

Arc Waveguides

For in- an out-coupling light to this resonator design, different coupling methods can be applied. As for silica disk-type microresonators, a taper fiber can be placed in the near vicinity of the resonator, or free beam excitation and collection via a microscope objective can be used. Furthermore, the applied laser writing lithography scheme also allows the controlled design of integrated polymer waveguide couplers. They can be structured with high precision such that evanescent coupling between waveguides and resonators can directly be achieved without further tuning. With this writing scheme, various waveguide designs can be implemented.

For the presented material platform, the so-called arc waveguides are very appealing. These are curved waveguides with both ends directly connected plane to the substrate. If a transparent glass slide is used, optical access through the backside of the slide is possible and can be performed by simple microscopy techniques. In the experiments, a microscope with a large field of view was used. By splitting the corresponding beam path for separating the incoming and outgoing directions, it was even possible to independently access both ends of the arc waveguides by a single objective. This highly simplifies the optical coupling process to these structures. Arc waveguides can be written with highest precision. Due to their short overall lengths, the produced arcs are mechanically stable and internal optical losses can be widely neglected.

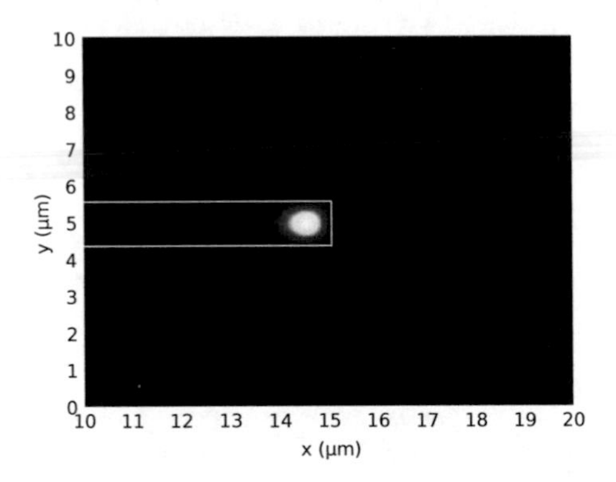

FIG. 6.3. Simulated mode profile for the fundamental TM mode in a polymer-based WGM microresonator (Photon Design Fimmwave). The disk has a diameter of 20 µm and a thickness of 1.2 µm. For the polymer, a refractive index of n = 1.5 at a wavelength of λ = 770 nm was assumed.

The end facets of the bended arc waveguides are directly structured on top of the silica slide. No additional cleaning or polishing steps have to be applied after production and the surface quality of the glass slide is directly transferred to the end facets of the arcs. Hence, even if the surface roughness of the polymer is high, the smooth interface between slide and polymer ensures good optical coupling without large amounts of scattering. An example of an arc waveguide can be seen in FIG. 6.2b. The shown structure has a diameter of 1.8 µm and an overall length of 40 µm. The gap between the arc and the resonator was chosen such that evanescent coupling between both structures can take place with high efficiency. Arc waveguides are a very elegant method for an efficient in- and out-coupling to integrated photonic structures. The arcs do not require a time consuming end-facet preparation as it is often mandatory in standard waveguide technologies. By evanescently coupling these arc waveguides to conventional planar waveguide designs, they can also be used as efficient external out-of-plane coupling elements for accessing structures reaching along the plane of the substrate slide.

6.4.3 Spectral Mode Analysis

For analyzing the mode structure of the polymer disks, light from a tunable external cavity diode laser (ECDL, see also Appendix B) emitting at a wavelength of 770 nm was coupled into the resonators by the evanescent fields of tapered optical fibers (see FIG. 6.4). The used ECDL system combines a small emission linewidth (300 kHz) with a large mode-hop-free scanning range (up to 15 nm). This allows examining the intrinsic mode structure of the excited whispering gallery modes (WGMs) as well as continuously scanning over more than two free spectral ranges (FSRs) of the system. By tuning the laser frequency over distinct WGMs, light is coupled into the resonators and the different modes can be observed as Lorentzian shaped dips in the transmitted laser power. The polarization of the incoming light was chosen such that resonance dips with maximum coupling depth were observed. For the normalization of the data sets, reference measurements were performed on uncoupled fiber tapers. The results of such measurements were then compared to the results in resonator coupled systems. This correlation technique gives direct access to the different resonance coupling depths and cancels out overlaying power modulations caused by slightly non-adiabatic taper transitions.

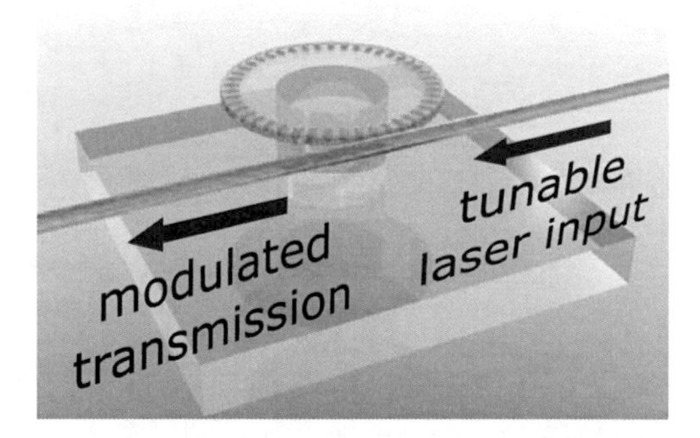

FIG. 6.4. A fiber taper is used to couple light from a tunable ECDL into polymer-based microresonators. While continuously sweeping the laser frequency, the transmitted light is modulated by the modes of the resonator (from [39]).

Q Factor

The Q factors of the observed WGMs were calculated from the transmission dips by a simple Lorentzian fit model. An additional linear term was added to match the local environment of the observed resonances. With this method, Q factors as high as $1.2 \cdot 10^4$ were measured at dip depths of up to 80 % (see FIG. 6.5). The fiber taper used for the experiments was not specifically phase matched with the modes of the resonator. The waist diameter of the fiber taper was 1.5 µm. All measurements were performed in full contact. By optimizing the coupling conditions of the combined system, it should be possible to observe even higher Q factors. Furthermore, it should also be possible to enhance the dip depths of the modes.

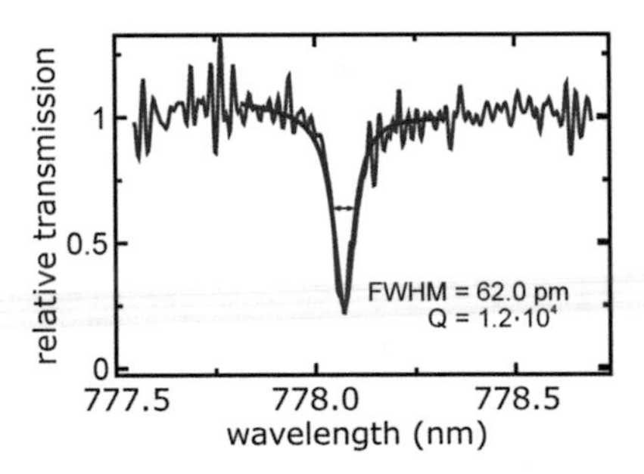

FIG. 6.5. High resolution scan of a WGM in a DLW polymer disk microresonator (diameter 20 µm, disk thickness 1.2 µm) coupled to a fiber taper with a waist diameter of 1.5 µm.

Mode Structure and FSR

By scanning over the available laser range, a clearly visible mode structure could be observed in correspondence to the FSR of the used microresonators (see FIG. 6.6). For wavelengths around 770 nm, a FSR of nearly 6.5 nm was measured. This value is in agreement with the theoretical value of 6.4 nm calculated from the optical path length. At a wavelength of 700 nm, the measured FSR decreased to 5.5 nm. This agrees as well with the analytically calculated value. Within one FSR, mainly four prominent modes can be observed. They belong to a single polarization state (TE or TM), but reflect different mode orders sharing the same azimuthal mode number. Thus, only modes up to the fourth order could be efficiently excited and observed by the used fiber taper.

Integrated System

The integrated polymer-based system, consisting of an easy accessible arc waveguide coupled to a WGM resonator, can also be examined by direct spectroscopy. This allows analyzing the mode structure over a much broader wavelength range, but this method is limited in their resolution by the availability of spectrometers. Exciting the WGMs by coupling a suitable white light source into one port of the arc allows determining the mode structure of a resonator by simply analyzing the transmission spectrum collected on the other side of the waveguide. However, it is also possible to use the intrinsic fluorescence of the diamond-dispersed polymer to spectrally resolve the mode structure of the coupled system. As the waveguides and the resonators are highly doped with nanodiamonds, the contained N-V defect centers can be spectrally excited by using green laser light ($\lambda = 532$ nm). Due to phonon sideband interactions, the defect centers emit a broad florescence spectrum quite similar to white light excitation.

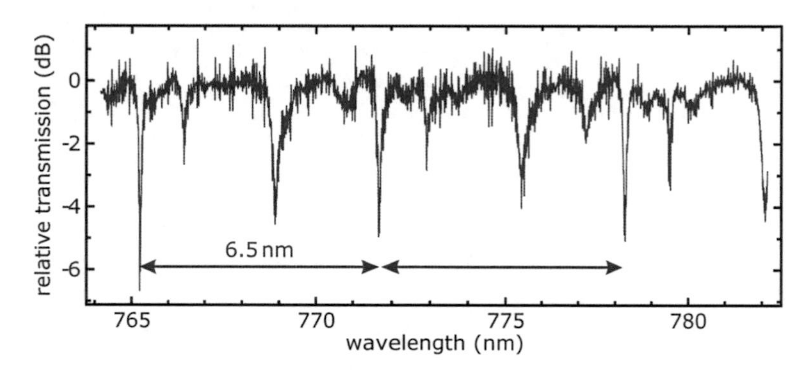

FIG. 6.6. Normalized transmission spectrum of a fiber taper which was evanescently coupled to a polymer-based WGM microresonator. The fiber taper had a waist diameter of 1.5 µm. The diameter of the resonator was 20 µm. The mode spectrum was observed by scanning the laser wavelength. The FSR is clearly visible and marked by the red arrows. The measured value of 6.5 nm is in good agreement with the analytic expectation based on the size of the resonator.

After coupling excitation light into one port of the arc waveguide, several independent emitters can be excited at once. The excitation light must not be resonant with the cavity. The fluorescence spectrum is generated inside the waveguide and can be used for further spectral analysis. Depending on the specific cavity resonance, parts of the fluorescence are coupled into the resonator. This opens an additional loss channel at those specific wavelengths and allows revealing the intrinsic mode properties of the resonant system by simply collecting the transmitted light at the second port of the arc waveguide. After filtering the excitation light, modulated background fluorescence spectra as shown in FIG. 6.7 can be observed by a grating spectrometer. The figure corresponds to the typical fluorescence spectrum of N-V color centers in diamond, but contains strong modulations caused by the WGMs of the coupled resonator system. A clear oscillatory behavior over the full spectral range can be observed. However, the underlying mode structure is not directly measurable from this spectrum.

For a more quantitative analysis of the background fluorescence within the polymer compound, an additional segmented Fourier transform was performed on the spectral data presented in FIG. 6.7. A direct Fourier analysis of the full data set was not possible and gives inconclusive results. The reason is that in microresonators with diameters below 20 μm the corresponding FSR is strongly wavelength dependent and differs substantially even from one azimuthal mode number to the next. Therefore, the full set was divided into 25 nm wide segments and several individual numerical fast Fourier transforms were performed on each of these sets. The different frequency amplitude ranges were then matched by normalizing the spectra of the different segments in relation to each other. Afterwards, a color coding scheme was applied to the resulting data. By merging the different wavelength ranges, the 3D color plot shown in FIG. 6.8 could finally be generated.

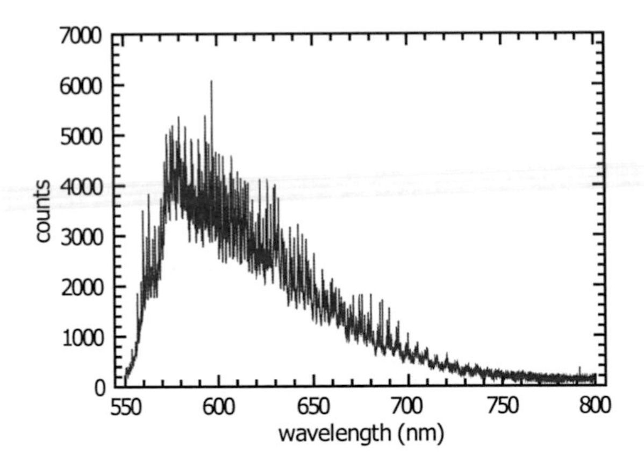

FIG. 6.7. Spectrum of the background fluorescence by exciting a nanodiamond-dispersed arc waveguide with 532 nm laser light at one port and collecting the filtered transmission on the other side. This spectrum is modulated by WGM resonances of an evanescently coupled microresonator. The laser power was below 1 mW to avoid thermal effects.

In this figure, the different Fourier components of WGMs inside the resonator are clearly visible as continuous bands evenly spaced by an integer multiple of the lowest observable Fourier frequency. This base frequency component is directly related to 1/FSR and corresponds well with the analytically estimated FSR values based on the actual size of the disc and the corresponding wavelengths. The black dotted lines are fits to these values and their different higher-order multiples. This result proofs the coupling between the directly structured arc waveguides and the resonators. It shows very good agreement between the observed modes and the spectral properties calculated from the geometrical size of the resonator.

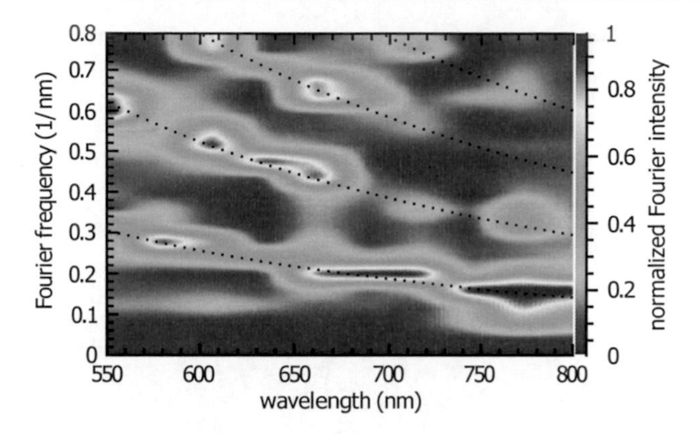

FIG. 6.8. Normalized segmented Fourier transform of the spectrum shown in FIG. 6.7. The lines in the intensity plot represent the different Fourier components. They are fits to the expected Fourier frequencies of the resonator modes and related to the wavelength dependent FSR of the system. The lowest frequency component relates to 1/FSR. Within the measured spectral range, a constant refractive index of $n = 1.5$ was assumed.

6.4.4 Outlook and Summary

The presented method of directly writing photonic circuits into nanodiamond-dispersed photopolymers offers the possibility for highly integrated chip-based quantum optical devices. The structuring technique is cost efficient and allows designing fully 3D structured photonic elements with nearly arbitrary shapes. In the last section, the basic material for the production of quantum-based optical elements and the required writing technique were presented. With this technique, the realizations of two key building blocks for quantum photonic circuitry were demonstrated. These are polymer-based WGM disk-type microresonators, and arc waveguides used as efficient coupling devices. Both elements contain quantum optical single photon emitters. The structures were analyzed for their specific optical properties. The general applicability of the nanodiamonds-dispersed polymer material for photonic microintegration could be demonstrated. The production process of the optical elements is scalable and allows the production of large amounts of devices for the realization of more complex designs.

The investigated polymer-based disk-type resonators are comparable to silica-based microresonators. Both designs show similar optical properties. Even though the observed Q factors were slightly lower as compared to the results measured in standard silica-based systems, by taking the higher absorption of the polymer material into account, the presented results are very promising for future applications. By using arc waveguides as evanescent couplers for such microresonators, a simple and efficient method of probing these structures could be demonstrated. The optical accessibility of the waveguide facets from below the substrate is an elegant method to circumvent classical coupling problems which are commonly encountered on nearly all chip-based designs. The perpendicular coupling approach is comparable to normal grating couplers which are often used to get access to planar chip-based waveguide structures. For these devices, the proper design of the grating is quite crucial and strongly depends on the required wavelength range. In an arc waveguide, total internal reflection is used to guide light and so the guiding mechanism is the same as for standard glass fibers. Hence, with arc couplers, the observed wavelength selectivity is less strict and efficient optical coupling is possible with high bandwidth.

It was not explicitly shown here, but a major advantage of the presented optical system is the possibility of incorporating and accessing individual single quantum emitters inside the structures. The active quantum functionality of waveguide-coupled 3D resonator systems could already be shown [39]. Single photon emission was realized within such structures and subsequent collection and routing through arc waveguides was successfully demonstrated. This is a first step to leverage integrated chip-based quantum optical technologies and it allows an extension to more complex designs. The applied dispersion technique is not specifically limited to a single photon system based on N-V defect centers. Nearly all stable quantum emitters in nanocrystalline form can be applied. The incorporation of other optical active elements, e.g., quantum dots or single molecules, should also be applicable. The utilized lithography method allows simple up-scaling of the principal design to highly complex integrated quantum systems where single photons are collected and routed through various optical elements, e.g., splitters, combiners, and spectral filters.

The applied structuring process is a highly flexible technique not limited to silica slides as substrate. It is applicable for a wide range of different other materials. Hence, even a hybrid integration of the presented polymer-based quantum optic elements with conventional optics in other material systems is possible. A direct integration of such quantum elements with lasers, detectors, other concurrent waveguide designs in nonlinear or electro-optic materials, or in microfluidic sensing devices are further promising features of polymer-based quantum photonics.

6.5 Two-Beam Direct Laser Writing Lithography

Another example for using photopolymers to fabricate novel photonic elements is 3D direct writing of waveguides with two-beam laser lithography. The method is based on the DLW process in diffusion-mediated photopolymers as introduced in Section 6.3.1. This lithography method allows the direct creation of inherently buried 3D waveguide structures without the requirement of additional development steps or capping layers. The two-beam DLW lithography was developed at the Technical University Berlin and is applied for patent [40].

For the two-beam DLW lithography, two Gaussian shaped laser beams are overlapped in their respective foci to create well localized dielectric waveguide structures inside a diffusion-mediated photopolymer. Non-trivial refractive index structures can be written by initiating intensity dependent polymerization and diffusion processes. The resulting intensity distribution of the writing beams was measured inside the photopolymer and a corresponding model for the refractive index was developed. The model is used to characterize the mode structure of written waveguides and to analyze the properties of these modes by means of numerical simulation software. In the following section, a short introduction to the two-beam DLW lithography is given. The results of numerical simulations are discussed and practical application examples for coupling polymer waveguides to standard telecom glass fibers are presented.

6.5.1 Writing Method

In diffusion-mediated photopolymers, refractive index contrasts are created between developed and undeveloped material portions. These volumes also define the core and cladding regions of the written waveguides. Within the dynamic range of the material, the achievable refractive index contrast mainly depends on the irradiated power and exposure time. Thus, the resulting refractive index profile is directly related to the absorbed total energy. For creating well defined index structures, a highly localized deposition is required. In contrast to the two-photon DLW process presented in the last section, in diffusion-mediated photopolymers localization can not be accomplished by simple focusing of a writing beam. Instead, special types of photoinitiators have to be used to achieve a non-linear material response. They can trigger the polymerization processes such that even the Gaussian energy distribution of a single focused laser beam is sufficient for defined process localization. Depending on the specific material response, different working regimes can be defined [41]. However, this technique lacks on the availability of suitable photopolymers exhibiting a large non-linear material response for such single laser beam excitations. Alternative writing techniques are required for the majority of linearly responding diffusion-mediated photopolymers.

Here, two focused Gaussian-shaped laser beams are spatially overlapped to create a localized intensity profile inside the material. A schematic of the corresponding setup is shown in FIG. 6.9. By crossing the beams in a common focus, a localized intensity maximum is created and X-shaped refractive index profiles can be achieved. The center region of the X contains twice the energy of the single laser beams. However, the arms of the X still remain at single beam energy level due to the angled superposition. The two-beam DLW lithography scheme highly increases the intensity localization in diffusion-mediated photopolymers as compared to approaches using single focused laser beams. Thus, even in photopolymers showing a negligible non-linear material response, well-defined deeply buried waveguide structures can be written.

Best waveguide shape with squarish cross-sections is achieved for a high angle of incidence between the two writing beams. However, in a physical system compromises between mechanical feasibility and optical aberrations are required. For the presented lithography setup, an incidence angle $\vartheta_{air} = 55°$ was chosen. This directly corresponds to an inner crossing angle $\vartheta_{polymer} = 33°$. In FIG. 6.10, the relations between these two different angles of incidence are shown.

The beam diameters in the overlapping region define the size of the written structures. Ideally, the two writing beams share the same size. In order to create small waveguide structures, both writing beams must be perfectly aligned to overlap in their respective foci. By moving the substrate, elongated refractive index structures can be created. Thereby, the size and shape of the structures can be controlled by simply varying the magnification of the writing lenses. It is also possible to change the position of the common focus in the depth of the material by moving the sample. Hence, arbitrary 3D waveguide trajectories can be produced and even non-planar optical structures can be realized with this method.

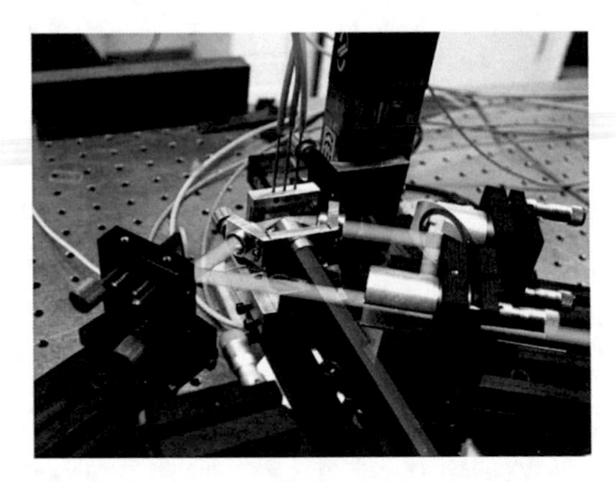

FIG. 6.9. Overlaid schematic of the two-beam direct laser writing setup for structuring waveguides inside diffusion-mediated photopolymers.

An epoxy-based SU-8 negative resist (micro resist technology) is used as basic component of the photopolymer. It contains a photoinitiator sensitive for wavelengths around 400 nm. At the operational wavelength of 850 nm, an average refractive index $n_{850nm} \approx 1.564$ was measured (Metricon 2010/M). For this wavelength, a maximum refractive index change in the order of $3 \cdot 10^{-3}$ is observed. The material response to irradiated energy is nearly linear which results in core refractive indices between 1.564 and 1.569.

The photopolymer is prepared inside a housing glass cuvette. The front side of the cuvette is covered by a 1 mm thick glass slide. The produced polymer film has a thickness of 300 µm. This enables full 3D structuring and allows a complete burying of the written structures. After finishing the writing processes, the photopolymer is bulk illuminated by UV light and hardened in a postbake process. Afterwards, the sample can be easily removed from the cuvette as stable solid device.

A drawback of the presented writing scheme is the possibility of spatial interferences between independent writing processes. The two writing beams are angled, thus unintended refractive index structures can be created by the subsequent superposition of previously independent beam paths. These effects have to be carefully taken in account by designing photonic structures with the two-beam DLW lithography.

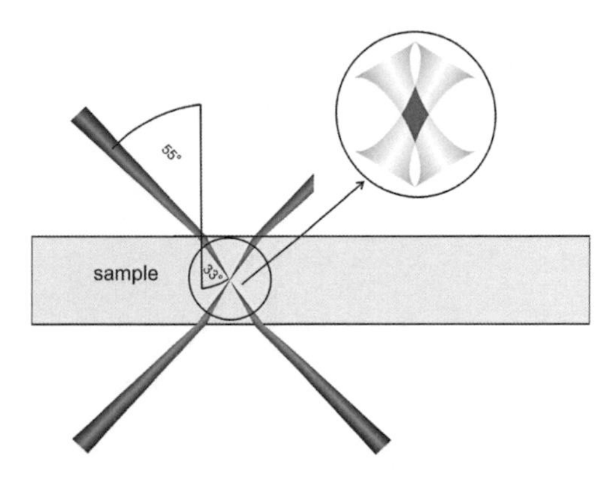

FIG. 6.10. Two-beam direct laser writing configuration. The inset shows the cross-section of the two intersecting laser beams.

By setting up the lithograph, proper alignment of the two writing beams is a crucial parameter. Therefore, a special sensor was build to mimic the conditions inside the photosensitive material. It can be used to determine the positions and intensity profiles of both writing beams by performing spatially resolved knife-edge scans. With this method, overlapping beams with spot sizes of less than 10 µm could be demonstrated.

6.5.2 Numerical Simulation

The obtained intensity distribution of the two writing beams was then transferred into a numerical simulation software specialized for waveguide design (Photon Design Fimmwave). Therefore, a corresponding model of the refractive index profile was developed. It assumes that the full dynamic range of the material can be used during processing. Based on the given linear material response of the used photopolymer, the resulting refractive index profile $n(x,y)$ is directly connected to the measured intensity profile $I(x,y)$ by

$$n(x,y)=\left(\frac{I(x,y)}{I_{\max}}\cdot\Delta n+1\right)\cdot n_0\,. \tag{6.2}$$

Here, n_0 and Δn are the refractive index and the maximum refractive index contrast.

For the simulations, a design wavelength of 850 nm (n_0 = 1.564 and Δn = 0.003) was chosen. The final mode structure of the written waveguides could then be calculated by using the two dimensional (2D) eigenmode analysis solver of the simulation software. Afterwards, the validity of the calculated numerical results was verified by analyzing the convergence properties of the applied model with other implemented numerical software solvers (FDTD, FEM).

6.5.3 Characterization

The different mode distributions for the calculated refractive index profile are shown in FIG. 6.11. They are based on waveguides written with beam diameters of 10 µm. The fundamental mode shows a symmetric Gaussian shape with a 1/e mode field diameter (MFD) of 12 µm. Due to the small refractive index contrast between the center and the arms of the X, higher-order modes slightly spread out along the corresponding crossing axes of the writing beams. Thus, small asymmetries between these writing beams result in an unequal distribution of modal energy (see FIG. 6.11f). The ability to separately control the optical properties of each writing beam is thus a highly crucial parameter not just for modeling but also for the reliable production of structured waveguides.

The mode distributions also show that the waveguide system exhibits a clearly ordered mode structure. Beside two fundamental polarization modes, higher order modes manifest themselves as multiple knots along the two crossing axes. Although the crossing between these axes is unequal due to the angled focusing, the different polarization modes seem to be degenerated within the simulations. It is not possible to discriminate between corresponding TE and TM modes of the same order by just their propagation constant. Therefore, a restriction of the analysis to a single polarization mode is sufficient.

In FIG. 6.12, the effective refractive indices for the first 7 TM modes are presented. For the lower order modes, the effective indices are all well separated. Thus, due to the different propagation constants and the highly dislocated field distributions, intermodal coupling should be negligible. The simulations, which have been performed with the propagation module of the applied software, verified this low transmission cross-talk between the different modes. It can be assumed that this feature also holds in reality under the assumption of sufficiently homogenous and symmetric writing conditions.

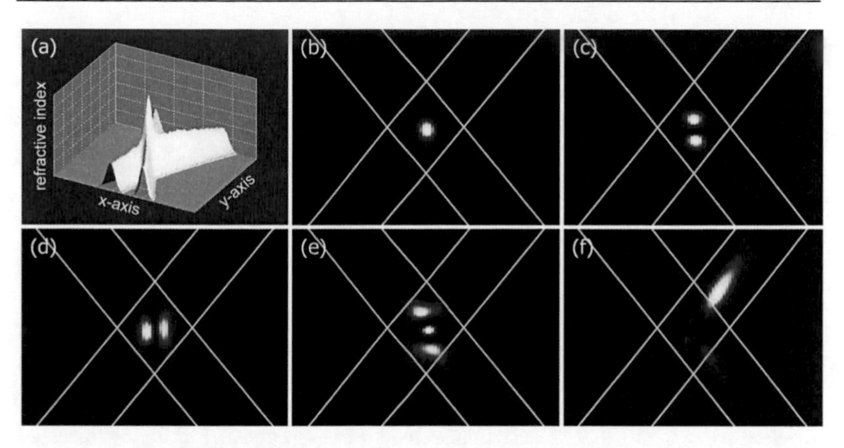

FIG. 6.11. (a) Refractive index profile of a waveguide written with writing beam diameters of 10 µm. The size of the depicted area is 240 µm x 135 µm. The refractive index ranges from 1.564 to 1.569. An operational wavelength of 850 nm was assumed. (b)-(f) Numerically calculated mode profiles for the measured refractive index distribution. TM modes up to the 5th-order are shown. The size of the profiles is 125 µm x 95 µm.

Due to the relatively low refractive index contrast between core and cladding regions, in diffusion-mediated photopolymers only weak guiding can be achieved for the buried structures. Thus, high bending losses may be observed in transmission. From numerical calculations, bending losses of more than 1 dB/cm were estimated for bends with radii below 500 µm. Thus, additional losses have to be carefully taken into account by designing curved buried waveguide structures with a low refractive index contrast.

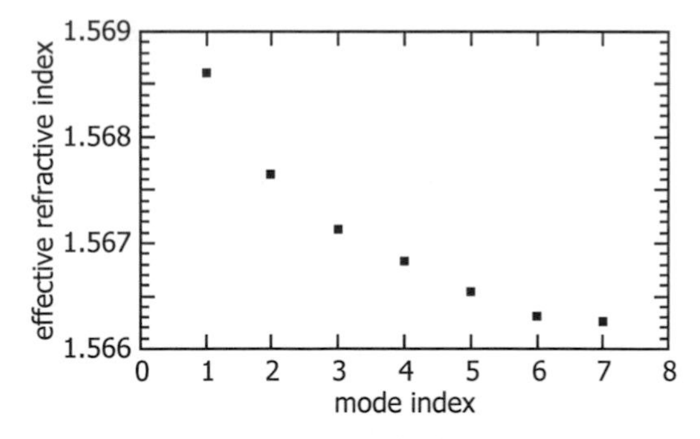

FIG. 6.12. Effective refractive indices of the first 7 TM modes. The shown simulation results correspond to the modes and conditions given in the caption of FIG. 6.11.

For characterizing the created refractive index structures, several parallel deeply buried waveguides were written in parallel. Thereby, the two writing beams had equal waist diameters of 10 µm. During writing, the intensities of the two beams were varied. From these measurements, an optimum writing energy density between 1500-2000 mJ/cm² was observed. After hardening and post-processing, the polymer sample was released from the cuvette. The two end facets of the polymer were carefully polished to optical quality (grain size 300 nm) and laser light at different wavelengths was then focused onto one of the facets.

The transmitted light was collected on the other facet with a glass fiber and the overall transmission losses were measured for different sample lengths via a cut-back method. The results of these measurements are shown in FIG. 6.13. The characterization setup only allows relative measurements and thus no explicit insertion loss can be extracted. Furthermore, the curves for the different wavelengths are not intensity correlated. From these data, an average waveguide transmission loss of 3 dB/cm was calculated. For the wavelength of 850 nm, this already includes material absorption and scattering losses in the range around 1.6 dB/cm. Other wavelength dependent losses were not specifically investigated.

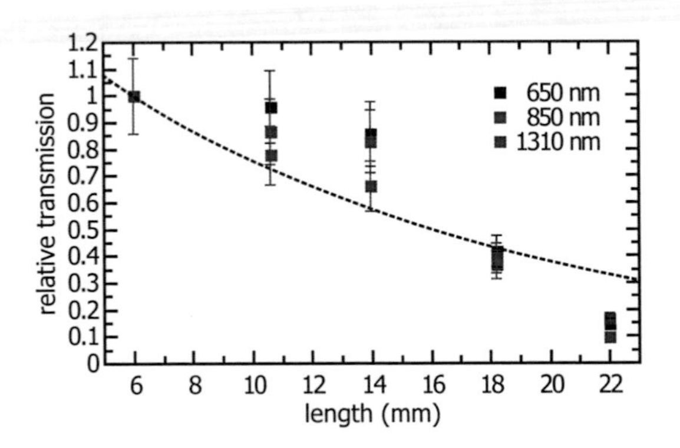

FIG. 6.13. Relative optical transmission vs. waveguide length. The different wavelengths are not intensity correlated. For all wavelengths, nearly identical curves were observed. The results are in agreement with an average waveguide transmission loss of 3 dB/cm (black line).

For experimentally analyzing the resulting mode structures, laser light at a wavelength of 850 nm was coupled into the waveguides and the intensity distributions at the output facet were imaged by a CCD camera. The measured intensity profiles are in good agreement with the results from the numerical simulations. In FIG. 6.14, an image of the output facet from a waveguide written with beam diameters of 10 µm is shown. The intensity distribution exhibits a clear Gaussian shape. For the depicted profile, a 1/e diameter of 12 µm was measured. This result is again in good agreement with the numerically calculated MFD of the fundamental mode in a waveguide of similar size and the waist diameter of the used writing beams.

The presented color plot also includes a fingerprint of the used two-beam writing configuration. This is illustrated by the two crossed black lines. However, it was also possible to observe completely symmetric Gaussian intensity profiles without such a fingerprint on some other waveguide structures. The measurements clearly show that waveguides with an X-shaped refractive index profile can be written by the two-beam DLW lithography. The measured mode profiles are in full agreement with the results of numerical simulations.

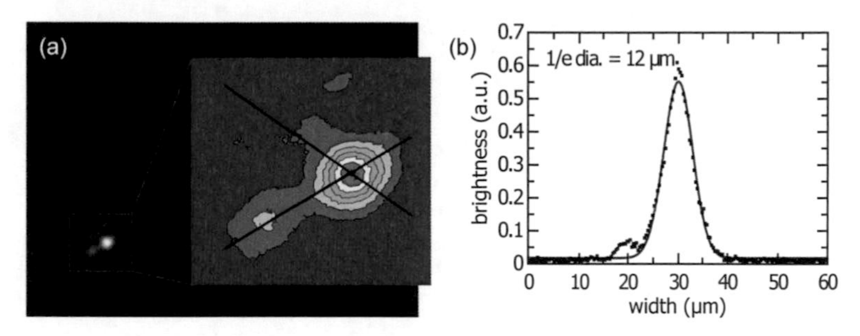

FIG. 6.14. (a) Intensity distribution at the output facet of a waveguide sample as imaged by a CCD camera. In the inset, a highly zoomed-in color plot is shown. The black lines represent the crossing of the two writing beams. (b) Intensity distribution along the lower black line. From a Gaussian fit, a 1/e mode field diameter of 12 µm is measured. The waveguide was written with 1/e beam waist diameters of 10 µm.

6.5.4 Fiber Pigtail Coupling

The actual setup utilizes focusing lenses with a numerical aperture $NA = 0.15$. With these lenses, writing beam diameters in a range between 5 µm and 60 µm can be realized. By using the mode field analysis as presented in the last section, the field diameters of the fundamental modes in waveguides written with different writing beam diameters can be calculated. For the lowest achievable beam size and at a design wavelength of 1310 nm, the two fundamental waveguide modes exhibit a mode field diameter (MFD) of nearly 10 µm. At this wavelength, the corresponding single mode cut-off diameter can also be calculated as 1.5 µm. In waveguides written with beam diameters below this size, solely the two fundamental TE and TM modes are guided. Although these diameters are not accessible with the actual writing optics, the calculations show that even with the actual range of writing beam sizes very high coupling efficiencies to MFD-fitted single mode optical glass fibers are achievable.

For demonstration, the coupling efficiency between the fundamental TM modes of a standard telecommunication single mode optical fiber (Corning SMF-28, core diameter d_{core} = 8.2 µm, core index n_{core} =1.4668, cladding index $n_{cladding}$ = 1.4624) and the presented polymer waveguides was calculated. It is assumed that the fiber is molded into the material and that the waveguide is directly connected to the core region of the fiber without spatial or angular mismatch. The coupling efficiency was calculated by varying the waveguide diameter and analyzing the coupling between the fundamental modes. The result of these simulations is shown in FIG. 6.15. For waveguides written with beam diameters of 4 µm, a maximum coupling efficiency of nearly 92 % can be achieved. However, the total slope of the efficiency curve is very flat. Even with beam diameters between 3 µm and 6 µm, coupling efficiencies of at least 90 % are observed.

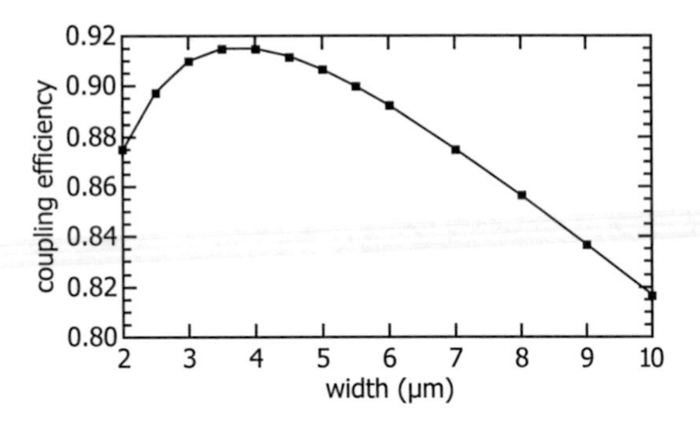

FIG. 6.15. Simulated coupling efficiency between the fundamental TM modes of a standard SMF-28 optical fiber and polymer waveguides written with different beam diameters.

6.5.5 Summary and Outlook

In the last section, the two-beam DLW lithography method for directly writing waveguides into diffusion-mediated photopolymers was presented and numerical investigations on the resulting waveguide structures were given. By measuring the transmission properties, the creation of optical waveguides with a minimum MFD of around 12 µm could be demonstrated. Thereby, it could be shown that the measured diameters of the waveguides are directly related to the waist diameters of the used writing beams. This shows the direct applicability of the writing scheme and verifies the locality of the written refractive index structures.

The two-beam lithography creates an intensity profile which directly transfers to an X-shaped refractive index distribution. By assuming a simple linear dependency between the writing intensities and the maximum refractive index contrast of the material, the creation of waveguides with clear mode structures could be demonstrated.

The fundamental mode of the system is mainly located in the center of the waveguide, while the higher order modes show an ordered distribution of their field maxima along the X-shaped axes. The intensity profiles of the different mode orders are highly dislocated and show a negligible intermodal overlap. The corresponding propagation constants are clearly distinguishable. It is possible to match the propagation constant and the mode area of the fundamental polymer waveguide mode to the fundamental mode of a SMF-28 standard telecommunication single mode glass fiber. Due to this matching, highest coupling efficiencies can be achieved between polymer waveguides and external optical elements.

The developed two-beam DLW technique is a simple and efficient solution for implementing a polymer-based lithographic scheme into the photonic industry. It provides a cost saving alternative to a multitude of classical connecting and coupling problems. In addition, the possibility of writing structures directly into photosensitive bulk materials may also open completely new fields of applications in polymer optics and photonics. The method is perfectly suited for the efficient production of photonic elements even in large volumes. By using a continuous writing scheme, also extended continuous waveguide structures can be fabricated in highest velocity. This is a major advantage over two-photon polymerization techniques in other materials. As these methods only allow slow sequential voxel by voxel writing of structures, the processing and manufacturing times are much higher especially for large and elongated elements. This substantially raises the production costs and limits the applicability of those techniques. In contrast, the two-beam DLW lithography method is a 3D-enabled technique which allows continuous writing of straight and curved waveguide structures with varying structure sizes. Furthermore, by simply changing the illumination and writing conditions, the size and shape of these structures can be dynamically controlled. The technique allows simple bulk processing and continuous writing of waveguides even with large core diameters. With this writing technique, also tapered or extended waveguides can be produced in high velocity and without introducing artificial surface roughness as commonly observed with voxel-based writing schemes.

Therefore, a possible application for this polymer-based technology is the flexible integration with other optical elements. An example is mode conversion between two types of glass fibers. With a tapered polymer waveguide, different modes of the fibers can be matched to each other even if the core sizes are different. By connecting the two fiber ends with such waveguide, the fundamental mode in the first fiber is efficiently coupled to the fundamental mode of the opposite fiber. After embedding both fiber ends into a polymer sample, a tapered waveguide can directly be written between the cores. With this technology, new realms of further applications are possible for solving a multitude of classical coupling and beam shaping problems in photonics. Especially the efficient coupling of laser light from diodes or VCSELs into glass fibers is a frequent problem in laser and telecommunication industries. Polymer-based systems offer a very appealing and cost efficient alternative to the up-to-date collimation and correction optics. These are mainly based on classical optical elements, e.g., lenses and mirrors, and thus complex and expensive alignment processes are often involved.

However, the flexible processing methods of polymers make this technology also very appealing for the microintegration with other heterogeneous waveguide systems. The photopolymer used for the two-beam DLW lithography is solid under normal ambient conditions, but can be softened by heating over a specific glass transition temperature. Within this state, the material can be casted or injection molded to nearly arbitrary shapes. After completely hardening the polymer, the samples can be separated from the preform and waveguides can be written on demand. However, as this material is mainly based on SU-8 photoresist, it firmly adheres to nearly all base coats. This also allows a simple and selective direct casting of the polymer onto various substrates.

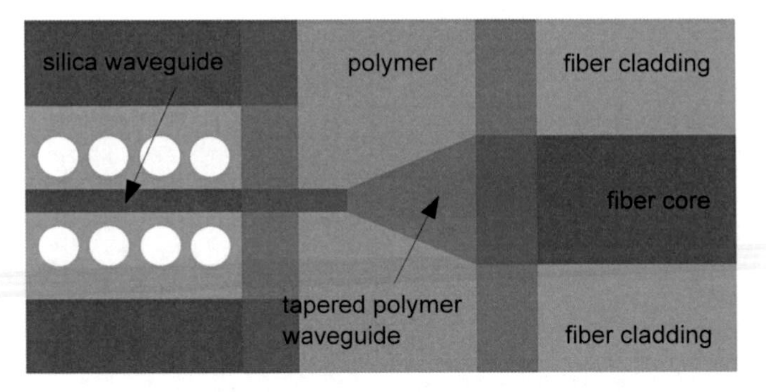

FIG. 6.16. Schematic of a photonic circuit partly embedded inside a polymer. Glass fibers can be embedded into the polymer to allow aligned waveguide structuring. Stable and reliable interconnects between fibers and ports can be realized. This highly simplifies coupling between integrated photonic elements and external components.

An interesting application would be the combination of such a casting method with the silica-based waveguide technology introduced in Chapter 5. A possible implementation of both technologies is illustrated in FIG. 6.16. Especially for more advanced chip designs including several hundreds of input and output ports, coupling to various external elements could be highly simplified by such a polymer-based intersection. A stable and permanent connection between several glass fibers and a microchip is still a challenging task in optical microintegration technology. By using polymer casting, the different glass fibers as well as the silica waveguides could be embedded into the material to create a sustainable permanent connection. For achieving a high coupling efficiency between silica waveguides and glass fibers, the two-beam DLW lithography offers a reliable method to directly write mode-matched polymer waveguides from one element to another. This way, it is also possible to individually adjust the waveguide trajectories to handle misalignments between the different optical components. Another option is to include beam transformation into the waveguide couplers. By tapering or extending the waveguide area, different mode field diameters can be matched and the coupling efficiency can be further optimized. By elongating the waveguides in one direction, the symmetry of the coupler can also be changed to allow better coupling between the rectangular profiles of the waveguides in the silica and the circular profile of the embedded glass fibers.

Simple waveguide structures and applications as versatile coupling elements are basic functionalities. By using two-beam DLW in diffusion-mediated photopolymers, also the integration of more complex photonic elements can be realized. Directional couplers, beam splitters and combiners, Fabry-Pérot interferometers, and other types of resonators can be directly written into the material. Also for such optical elements, the major advantage of photopolymers as basic material is the simple and cost efficient production process in contrast to the elaborate and expensive lithographic methods used for structuring in other material systems.

A further possibility for the realization of additional optical functionalities in this material is the combination of directly written waveguide structures with holographic volume gratings. They can be created inside the photopolymer by nearly the same processes. As in standard two-beam laser writing, two intersecting laser beams are overlapped inside the material. However, by using coherent laser beams coming from a single source, a stable interference pattern can be generated inside the photopolymer. A corresponding refractive index contrast between the different grating layers is then created by diffusion in analogy to the waveguide formation process. By taking the available contrast range of the material into account, these wavelength selective gratings can be combined with waveguides written by the two-beam DLW lithography method. This allows the integration of spectral filtering inside the waveguides and also wavelength selective routing into different channels can be realized. If the silica-based waveguide system presented in Chapter 5 is used for quantum optics experiments, silica-attached single-photon emitters in nanodiamonds could be pumped via connected optical glass fibers. Integrated pumping via fibers is often not possible as the occurrence of strong background fluorescence prevents the detection of single photons coming from the emitters. By using polymer-based holographic spectral filters, the spectral components of the fluorescence can be filtered out before the pump light is coupled into the pure silica of the waveguides.

Another asset of polymer-based optical elements is their simple miscibility with other materials. In analogy to the doping processes in conventional solid state materials, photopolymers can simply be mixed with other components, e.g., active materials like single photon emitters, and organic or inorganic complexes. By adding dye molecules or quantum dots to the basic photopolymer composition, the realization of optical amplifiers or wavelength shifters is possible. Another example is given by adding liquid crystals to the material. Actively switchable optical components and circuits can be realized and have been successfully demonstrated [37]. However, by using the two-beam DLW lithography, these elements can be continuously written as already capped bulk systems integrating couplers and connectors to their optical functionality.

[1] A. Yeniay, R. Gao, and K. Takayama, "Ultra-low-loss polymer waveguides," *J. Light. Technol.*, vol. 22, no. 1, pp. 154–158, 2004.

[2] K. Shimomura, T. Aizawa, N. Tanaka, and S. Arai, "Polymer Optical Waveguide Switch Using Thermo-Optic Total-Internal-Reflection and Strain-Effect," *Photonics Technol. Lett.*, vol. 4, no. 4, pp. 197–199, 2010.

[3] B. Lee, C. Lin, X. Wang, R. T. Chen, J. Luo, and A. K. Y. Jen, "Bias-free electro-optic polymer-based two-section Y-branch waveguide modulator with 22 dB linearity enhancement," *Opt. Lett.*, vol. 34, no. 21, pp. 3277–3279, 2009.

[4] S. Singh, V. R. Kanetkar, G. Sridhar, V. Muthuswamy, and K. Raja, "Solid-state polymeric dye lasers," *J. Lumin.*, vol. 101, no. 4, pp. 285–291, 2003.

[5] B. H. Ma, A. K. Jen, and L. R. Dalton, "Polymer-Based Optical Waveguides: Materials, Processing, and Devices," *Adv. Mater.*, vol. 14, no. 19, pp. 1339–1365, 2002.

[6] K.-L. Deng, T. Gorczyca, B. K. Lee, H. Xia, R. Guida, and T. Karras, "Self-Aligned Single-Mode Polymer Waveguide Interconnections for Efficient Chip-to-Chip Optical Coupling," *J. Sel. Top. Quantum Electron.*, vol. 12, no. 5, pp. 923–930, 2006.

[7] S. Kalluri, M. Ziari, V. Chuyanov, W. H. Steier, D. Chen, B. Jalali, H. Fetterman, and L. R. Dalton, "Monolithic Integration of Waveguide Polymer Electrooptic Modulators on VLSI Circuitry," *Photonics Technol. Lett.*, vol. 8, no. 5, pp. 644–646, 1996.

[8] J. Wang, P. J. Shustack, and S. M. Garner, "High Performance Polymer Waveguide Devices via Low Cost Direct Photolithography Process," in *Proceedings of SPIE*, 2002, vol. 4904, pp. 129–138.

[9] S. Kawata and H.-B. Sun, "Two-photon photopolymerization as a tool for making micro-devices," *Appl. Surf. Sci.*, vol. 208, pp. 153–158, 2003.

[10] A. Barsella, H. Dorkenoo, and L. Mager, "Near infrared two-photon self-confinement in photopolymers for light induced self-written waveguides fabrication," *Appl. Phys. Lett.*, vol. 100, no. 22, p. 221102, 2012.

[11] C. Ye, K. T. Kamysiak, A. C. Sullivan, and R. R. McLeod, "Mode profile imaging and loss measurement for uniform and tapered single-mode 3D waveguides in diffusive photopolymer," *Opt. Express*, vol. 20, no. 6, pp. 6575–6583, 2012.

[12] O. Sugihara, H. Tsuchie, H. Endo, N. Okamoto, T. Yamashita, M. Kagami, and T. Kaino, "Light-Induced Self-Written Polymeric Optical Waveguides for Single-Mode Propagation and for Optical Interconnections," *Photonics Technol. Lett.*, vol. 16, no. 3, pp. 804–806, 2004.

[13] K. J. Vahala, Ed., "Optical Microcavities," in *Advanced Series in Applied Physics*, vol. 5, Singapore: World Scientic, 2004.

[14] G. Hougham, G. Tesoro, and A. Viehbeck, “Influence of free volume change on the relative permittivity and refractive index in fluoropolyimides,” *Macromolecules*, vol. 29, no. 10, pp. 3453–3456, 1996.

[15] A. J. Beuhler, D. A. Wargowski, T. C. Kowalczyk, and K. D. Singer, “Optical polyimides for single-mode waveguides,” in *Proceedings of SPIE*, 1993, vol. 1849, pp. 92–103.

[16] C. P. Wong, Ed., *Polymers for electronic and photonic applications*. New York: Academic Press, 1993.

[17] S. Herminghaus, D. Boese, D. Y. Yoon, and B. A. Smith, “Large anisotropy in optical properties of thin polyimide films of poly(p-phenylene biphenyltetracarboximide),” *Appl. Phys. Lett.*, vol. 59, no. 9, pp. 1043–1045, 1991.

[18] Z. Zhang, P. Zhao, P. Lin, and F. Sun, “Thermo-optic coefficients of polymers for optical waveguide applications,” *Polymer (Guildf).*, vol. 47, no. 14, pp. 4893–4896, 2006.

[19] J. Mark, *Physical properties of polymers handbook*. New York: Springer, 2007.

[20] L. A. Eldada, S. Yin, R. A. Norwood, and J. T. Yardley, “Affordable WDM components: the polymer solution,” in *Proceedings of SPIE*, 1998, vol. 3234, pp. 161–174.

[21] R. A. Norwood, R. Gao, J. Sharma, and C. C. Teng, “Sources of loss in single-mode polymer optical waveguides,” in *Proceedings of SPIE*, 2001, vol. 4439, pp. 19–28.

[22] J.-L. Goudard, P. Berthier, X. Boddaert, D. Laffitte, and J. Périnet, “New qualification approach for optoelectronic components,” *Microelectron. Reliab.*, vol. 42, no. 9–11, pp. 1307–1310, 2002.

[23] D. Blitz and B. J. Pernick, “Polarization properties of photopolymers for use in holographic and coherent optical systems,” *Appl. Opt.*, vol. 32, no. 32, pp. 6501–6502, 1993.

[24] Y. Xia, E. Kim, and G. M. Whitesides, “Micromolding of Polymers in Capillaries: Applications in Microfabrication,” *Chem. Mater.*, vol. 8, no. 7, pp. 1558–1567, 1996.

[25] J. A. Rogers and R. G. Nuzzo, “Recent progress in soft lithography,” *Mater. Today*, vol. 8, no. 2, pp. 50–56, 2005.

[26] C. Choi, “Fabrication of optical waveguides in thermosetting polymers using hot embossing,” *J. Micromechanics Microengineering*, vol. 14, no. 7, pp. 945–949, 2004.

[27] T. Han, S. Madden, M. Zhang, R. Charters, and B. Luther-Davies, “Low cost nanoimprinted polymer waveguides,” in *Conference on Optoelectronic and Microelectronic Materials and Devices (COMMAD)*, 2008, pp. 185–188.

[28] L. Eldada, C. Xu, K. M. T. Stengel, L. W. Shacklette, and J. T. Yardley, "Laser-fabricated low-loss single-mode raised-rib waveguiding devices in polymers," *J. Light. Technol.*, vol. 14, no. 7, pp. 1704–1713, 1996.

[29] R. Yoshimura, M. Hikita, S. Tomaru, and S. Imamura, "Low-loss polymeric optical waveguides fabricated with deuterated polyfluoromethacrylate," *J. Light. Technol.*, vol. 16, no. 6, pp. 1030–1037, 1998.

[30] D. Tomić and A. Mickelson, "Photobleaching for Optical Waveguide Formation in a Guest-Host Polyimide," *Appl. Opt.*, vol. 38, no. 18, pp. 3893–3903, 1999.

[31] M.-C. Oh, S.-Y. Shin, W.-Y. Hwang, and J.-J. Kim, "Poling-induced waveguide polarizers in electrooptic polymers," *Photonics Technol. Lett.*, vol. 8, no. 3, pp. 375–377, 1996.

[32] Y. Huang, G. Paloczi, A. Yariv, C. Zhang, and L. R. Dalton, "Fabrication and replication of polymer integrated optical devices using electron-beam lithography and soft lithography," *J. Phys. Chem. B*, vol. 108, no. 25, pp. 8606–8613, 2004.

[33] M. Deubel, G. von Freymann, M. Wegener, S. Pereira, K. Busch, and C. M. Soukoulis, "Direct laser writing of three-dimensional photonic-crystal templates for telecommunications," *Nat. Mater.*, vol. 3, no. 7, pp. 444–447, 2004.

[34] G. von Freymann, A. Ledermann, M. Thiel, I. Staude, S. Essig, K. Busch, and M. Wegener, "Three-Dimensional Nanostructures for Photonics," *Adv. Funct. Mater.*, vol. 20, no. 7, pp. 1038–1052, 2010.

[35] L. Li, R. Gattass, E. Gershgoren, H. Hwang, and J. Fourkas, "Achieving λ/20 resolution by one-color initiation and deactivation of polymerization," *Science (80-.).*, vol. 324, no. 5929, pp. 910–913, 2009.

[36] A. C. Sullivan, "Tomographic Characterization of Volume photopolymers for integrated optics," University of Colorado, 2008.

[37] G. M. N. A. Hassanein, "Characterization of polymer dispersed liquid crystal for photonic device applications," City University of Hong Kong, 2012.

[38] A. Al-Ghamdi and E. Mahrous, "Dye-Doped Polymer Laser Prepared by a Novel Laser Polymerization Method," *Int. J. Electrochem. Sci.*, vol. 6, pp. 5510–5520, 2011.

[39] A. W. Schell, J. Kaschke, J. Fischer, R. Henze, J. Wolters, M. Wegener, and O. Benson, "Three-dimensional quantum photonic elements based on single nitrogen vacancy-centres in laser-written microstructures," *Sci. Rep.*, vol. 3, p. 1577, 2013.

[40] J. Gortner, S. Orlic, M. Seifried, and C. Stark, "Verfahren zum Herstellen eines Lichtwellenleiters in einem Polymer," DE 10 2011 017 329 A1, 2011.

[41] A. C. Sullivan, M. W. Grabowski, and R. R. McLeod, "Three-dimensional direct-write lithography into photopolymer," *Appl. Opt.*, vol. 46, no. 3, pp. 295–301, 2007.

Chapter 7

Alternative Waveguide and Coupler Designs

7.1 Introduction

The structures discussed in the previous chapters are based on dielectric waveguiding. They were made of silica-on-silicon or optical polymers. In silica, exceptional optical properties can be achieved for applications in the visible (VIS) and near-infrared (NIR) spectral ranges. Furthermore, the highly developed processing techniques from the semiconductor industry can be applied. Polymer materials in contrast typically show significantly higher losses; however, they allow a greater flexibility in the available processing techniques and for designing their optical properties.

Several other material systems can be used for integrated quantum photonics. A short summary can be found in the introduction to Chapter 5. A very interesting alternative to silica is silicon nitride (Si_3N_4) [1], [2]. The material is transparent in the visible (VIS) spectral range, but optimum transmission is observed in the mid- and near-infrared regions [3]. Silicon nitride layers can be chemically deposited on various substrates and standard processing techniques can be applied. A high index of refraction (n = 2.016, λ = 589 nm) allows high density waveguide integration for photonic circuits working in the VIS spectral range [4]. These waveguide structures can be cladded by silica or other compatible materials with lower indices of refraction. In contrast to silica, the structure sizes can be highly reduced and even sharp 90° waveguide bends can be achieved with low optical losses. All these possibilities are highly interesting for applications in integrated quantum optics. By using high index waveguide materials, strong field confinement can be achieved. This increases the possible interaction with attached photon emitters and allows higher levels of integration. Hence, optical losses and the influence of decoherence can be reduced with such materials.

In many applications, even higher levels of integration are necessary. For the design of photonic circuits, smallest structure sizes in the range of nanometers are required to be competitive to standard microelectronic circuits. But within dielectric waveguides, the maximum field confinement and thus the minimum structure sizes are limited by roughly the wavelength of the transmitted light. By using waveguide materials with a higher index of refraction, a small but limited decrease in size can be achieved. An alternative waveguide technology is based on nanoplasmonics in conductive media [5]. In these nanostructured systems, the optical energy of the guided fields is concentrated in so-called surface plasmon polaritons (SPPs). With such elements, minimum structure sizes of around 10-20 nm can be realized for applications in the visible spectral range. As those waveguide structures maintain the quantum properties of single photons, they are possible candidates for future integrated quantum processing devices [6]. The realization of a single plasmon transistor could also be demonstrated [7].

In the following chapter, a short introduction to both alternative waveguide designs is given. As silicon nitride processing is similar to silica processing, merely the most general aspects are summarized. For nanoplasmonics, the physics of SPPs is briefly presented and fundamental waveguide properties are discussed. Afterwards, a novel coupler design for the efficient coupling of light between dielectric and plasmonic waveguides is introduced. The presented coupler design is applied for patent [8]. The results of further investigations on that system are published elsewhere [9].

7.2 Silicon Nitride Waveguides

Silicon nitride is another silicon-based chemical compound. It can be produced by heating silicon powder up to 1500 °C in a nitrogen atmosphere. Nitride films on silicon wafers are commonly fabricated by various chemical vapor deposition (CVD) techniques. They can be performed at high or low gas pressures (HPCVD or LPCVD). Most commonly used are plasma enhanced processes (PECVD) allowing a hybrid deposition of SiO_x and SiN_x films. By varying the processing parameters, different types of nitrides with slightly different optical and mechanical properties can be produced. Thereby, the major difference is due to modifications in the density of the deposited films. This parameter directly relates to differences in the refractive index and the absorption properties of the materials [2].

The processing of silicon nitride films is similar to the processing of silica. Standard lithography techniques from semiconductor industry can be applied. The material allows well-defined structuring with highest resolutions by using e-beam lithography. The higher index of refraction in nitrides often requires an increased resolution for the structuring process. By comparing to silica, a size reduction of around 0.75 is normally required. This condition also holds for structuring the gaps in directional or other coupler designs. The decreased tolerances are setting high demands for the lithographic structuring processes, but allow higher densification compared to silica and other low refractive index materials.

The high refractive index of silicon nitride allows the direct patterning of waveguides onto a wide range of substrate materials. Antiresonant spacer layers or suspended waveguide designs are not required when substrates with a lower index of refraction are used. Silicon nitride can be deposited on silica which is ideally suited for waveguide applications in quantum optics [10]. Due to the low optical loss in silica, standard rectangular silicon nitride channel waveguides can directly be patterned on top of such silica substrates without a spacer. Silica can also be used for capping waveguides in buried channel designs. For example, such waveguides are extensively investigated by Bauters *et al.* [1]. Ultra-low loss waveguides (λ = 1550 nm) with 8-9 dB/m for bends with 0.5 mm bend radii could be successfully demonstrated in LPCVD nitride films.

7.3 Plasmonic Waveguides

Electromagnetic waves are normally guided within dielectric structures. When metals are applied, a different guiding mechanism is involved. It is carried out by surface plasmon polaritons (SPPs) which can be used for plasmonic waveguiding. A strong localization of the fields inside such elements allows the design of highly integrated waveguide structures with field densities not achievable within conventional dielectric structures. In the following section, a short introduction to SPPs is given and plasmonic waveguide designs are discussed. A more detailed summary can be found in [11].

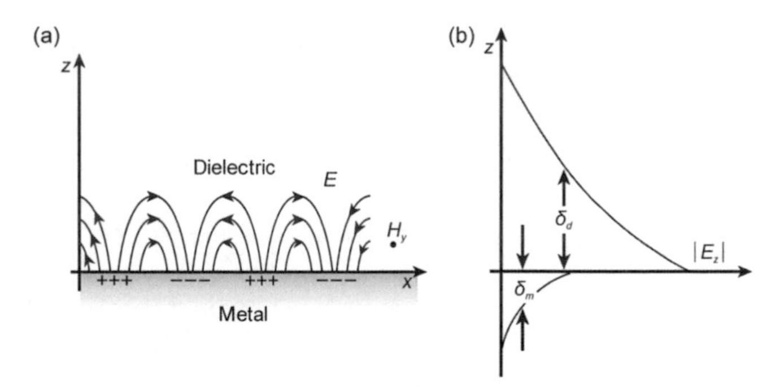

FIG. 7.1. Surface plasmon polariton (from [5]). (a) Field configuration along the interface between dielectric and metal. (b) The perpendicular E_z field decays exponentially and can be described by two penetration depths δ_d and δ_m inside dielectric and metal, respectively.

At the interface between a conductor and a dielectric, coupling can occur between an electromagnetic excitation and the electron plasma inside the conductor. This leads to the formation of self-sustained propagating electromagnetic surface waves where electrical charges at the surfaces of the conductor collectively oscillate in a hybridized excitation with the incoming photons (see FIG. 7.1). These coupled oscillations can be fully described by Maxwell's equations. By combing the dielectric and plasmonic properties at both sides of an interface, a common wave equation can be derived.

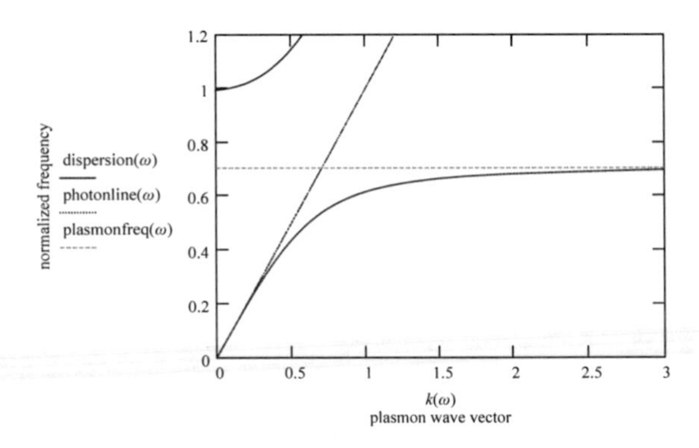

FIG. 7.2. Dispersion relation of SPPs at a silver/air interface based on the Drude approximation (ω_p = $1.2 \cdot 10^{16}$ rad/s, γ = $1.45 \cdot 10^{13}$ Hz, $\varepsilon_{dielectric}$ = 1). The photon line corresponds to the dispersion relation of light in free space.

The dispersion and other optical properties in metals can be found from the dielectric function of a free electron gas (Drude model) [12]. At the frequency ω, this function is given by the plasma frequency ω_p and the scattering rate γ, which accounts for the dissipation of the electronic motion, by

$$\varepsilon_{metal}(\omega) = 1 - \frac{\omega_p^2}{\omega^2 - i\gamma\omega}. \tag{7.1}$$

With the dielectric constant $\varepsilon_{dielectric}$ of the dielectric, the wave equations on both sides of the interface can be matched. By solving the resulting set of equations, the optical properties of SPPs are calculated. It can be found that, in contrast to ordinary dielectric waveguides, SSPs only exist in TM polarized states [13].

In most cases, single flat interfaces between dielectrics and conductors are investigated. By choosing appropriate boundary conditions and using the two dielectric functions $\varepsilon_{dielectric}$ and ε_{metal}, the dispersion relation between the frequency and the in-plane wave vector of SPPs can be found as [12]

$$k_{SPP}(\omega) = \frac{\omega}{c} \cdot \sqrt{\frac{\varepsilon_{dielectric} \cdot \varepsilon_{metal}(\omega)}{\varepsilon_{dielectric} + \varepsilon_{metal}(\omega)}} \,. \tag{7.2}$$

In FIG. 7.2, the typical dispersion relation for single interface geometries is exemplarily shown for air as dielectric and a simple Drude approximation for silver as metal. Due to the bound nature of SPPs, the curve stays well below the corresponding photon line for the dispersion of light in free space. Thus, phase matching by a grating or other coupler device is always required for the efficient coupling between dielectric and plasmonic excitations [14]. In real metals, electronic interband transitions are limiting the applicability of the Drude model to the visible and higher optical frequencies [12]. The dispersion relation of metals also shows a finite limit for the wave vector at the so-called surface plasmon frequency ω_{sp}. This results in a lower bound for the possible wavelengths of SPPs inside plasmonic waveguides.

At the interface, SPPs around the specific plasmon frequency of the metal show very high field confinements. However, due to an increased attenuation at this frequency, ultra-short propagation lengths are typically observed. At frequencies close to ω_{sp}, the plasmon field can be confined to roughly half of the wavelength in the dielectric. Thereby, at visible frequencies, typical propagation lengths in the lower micrometer range can be realized. A typical extent for fields inside the metal is around 20 nm. This value is constant over a wide frequency range. Due to their high field confinements, SPPs are perfectly suited for applications in advanced quantum optics [5]. They allow an enhanced interaction between the plasmon fields and, external quantum emitters or other sources of non-classical light. Furthermore, a reduction in the overall size of integrated quantum systems can be realized for achieving higher levels of environmental isolation. Therefore, choosing a good trade-off between strong field confinement and a sufficient propagation length is most crucial to a design.

Another often used geometry for observing SPPs is based on multilayer sandwiches often referred to as insulator/metal/insulator (IMI) or metal/insulator/metal (MIM) heterostructures [15]. If the separation between the interfaces is below the decay length of the interface modes, the separate SPPs couple to each other. Hence, they can be described as a combined system of single interface SPPs. In IMI geometries with thin metal structures, attenuation can be reduced and long-range SPPs can be observed [16]. On the other hand, in MIM structures, the thickness of the insulator can be chosen such that strong field localization can even be achieved for frequencies far from ω_{sp}. This allows a geometrical tuning of the plasmon properties [17]. If air is used at the insulator section, ultra-low losses can be achieved. The design is then referred to as a plasmonic slot waveguide [15].

During the recent years, both geometries could be successfully applied to demonstrate plasmonic waveguiding. Most common are all types of metallic strip waveguides, but also slot waveguides are frequently used [6]. Due to the three dimensional structure of these designs, often several metal/insulator interfaces to the substrate or other close-by dielectrics have to be considered. The individual SPPs can sum up to a coupled SPP system if the geometrical ratios allow for at least some interactions between the different fields. For chip-based waveguide applications, long-range SPPs with ultra-thin metals are interesting. Here, the fields show a strong extent into the low-loss dielectric environment and very long propagation lengths can be achieved. In such structures, Gaussian shaped modes without a cut-off frequency are observed. However, if highest field confinement is necessary, tight field localization to the metal surface is required and thus high transmission losses are involved. For sub-wavelength mode confinement, the propagation lengths are limited to the micron and sub-micron range [6].

The production of metallic stripes is fully compatible to standard electronic clean room processing. Metallic layers can be deposited on various substrates by using different vapor deposition or spraying techniques. The structuring of metal films is similar to semiconductor processing. Due to the reduced structure size, an increased structuring resolution is required and direct processing via e-beam or by imprinting techniques is preferred. Depending on the material, wet chemical etching can often be applied to selectively remove metallic layers. For plasmonic waveguide designs, silver and gold are used as metals due to their good mechanical and optical properties. However, thin silver films tend to strongly oxidize in air and often require additional protection.

7.4 Photon-to-Plasmon Coupler

In the previous section, alternative waveguide designs for increasing the density of chip-integrated applications were presented. In contrast to pure silica-based systems, higher optical losses are observed within the VIS and NIR spectral ranges. However, for chip-based applications, silicon nitride is preferable wherever higher levels of integration and better material compatibility are required. As shown in the previous section, highest levels of optical integration can be achieved with plasmonic waveguide structures. As direct coupling to plasmons is difficult, efficient methods for a controlled in- and out-coupling to and from plasmonic waveguides have to be developed [18].

In most up-to-date plasmonic applications, grating couplers working via simple photon scattering are used to convert free space photons into plasmons and vice versa [19]. A disadvantage of this method is the requirement for an angled alignment between the different optical elements. Grating couplers often work out-of-plane which highly complicates the coupling between different planar structures. The high optical losses of plasmonic waveguides limit their possible applications to very short distances. Thus, by using grating couplers, a highly dense spacing of coupling elements is required. This condition complicates the applicability of grating couplers even more and limits their application to foremost simple experimental designs with a low number of accessible optical ports.

Another option for solving the coupling problem is given by combining different planar waveguide designs to allow bridging the large distances on a microchip. Thereby, the combination of silicon nitride with plasmonic waveguides is perfectly suited for a highly integrated platform technology in advanced quantum optics and plasmonics. Both materials can directly be structured on a common substrate and allow hybrid integration with short and long ranging transmission lines. Even the connection of separate plasmonic circuits, all located on different positions of a single microchip, becomes possible by using such a hybrid technology. For such integrated designs, highly efficient interconnections between guided photons and plasmons are required. The overall coupling efficiency can be optimized by choosing appropriate plasmonic and photonic waveguide structures.

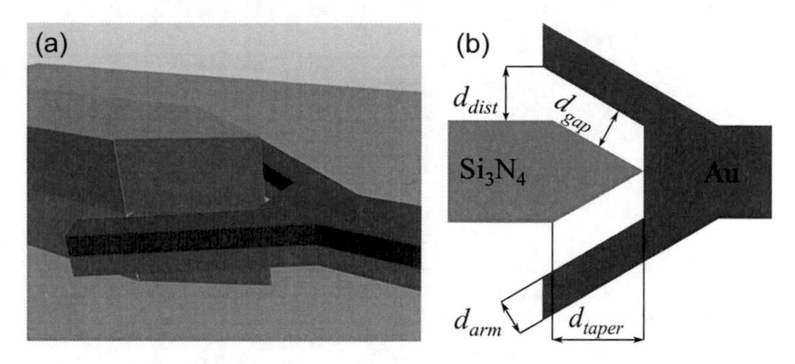

FIG. 7.3. Photon-to-plasmon coupler (from [9]). (a) Illustration of the structure. (b) Top view of the coupler with four free design parameters (gap distance d_{gap}, taper length d_{taper}, arm thickness d_{arm}, arm distance d_{dist}).

During the last decade, a variety of possible photon-to-plasmon coupler designs were proposed and investigated. This mainly includes numerical investigations performed in two [17], [20] and three [21]–[25] dimensions (2D and 3D). However, many of the presented designs are not suitable for applications in quantum optics or general hybrid integration. They regularly require specific waveguide designs and are often not compatible with the standard lithographic processes normally used in laboratories and semiconductor fabrication plants. For quantum optics applications, free access to the guided fields is required. Thus, designs with buried waveguides [24], [25] can not be adapted. Furthermore, many of the numerical proposals which are stating highest coupling efficiencies are only based on 2D calculations and assume infinitely thick materials [17], [20]. Transferring such structures to real designs is often not possible. By using purely analytic approaches, like the coupled mode theory (CMT) [25], the coupling efficiencies are often overestimated [26]. Hence, ideal coupler designs combine easy fabrication processes in standard lithography with simple structures suitable for a full numeric or analytic analysis in 3D.

An example of such a hybrid structure is presented in FIG. 7.3. The design was developed with respect to fabrication constraints for quantum optics applications in the VIS spectral range. It consists of a dielectric silicon nitride waveguide and a plasmonic waveguide based on gold. Both waveguide are produced as channel waveguides on a silica substrate. The dielectric part has a width of 510 nm and a height of 200 nm. The plasmonic part is made from gold with identical waveguide width but a reduced structure height of 50 nm. The dimensions of the dielectric waveguide were optimized to allow single mode operation at wavelengths around 780 nm. Afterwards, the dimensions of the plasmonic channel were matched to the effective index of the dielectric part.

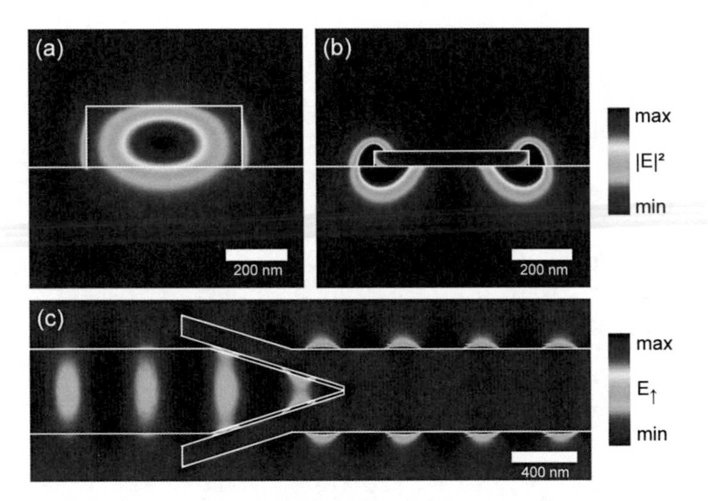

FIG. 7.4. Electric energy distribution $|E|^2$ inside (a) dielectric and (b) plasmonic waveguides. The maximum field density between both distributions shows a 1:4 field enhancement in the plasmonic part. (c) Top view of the complete coupler structure. The electric field strength at the interface between the waveguides and the silica substrate is shown (from [9]).

The coupling region between both waveguides is structured by a V-shaped directional coupler allowing an adiabatic transition between the two fields. Due to the high transmission loss in the plasmonic section of the converter, a trade-off between the effective coupling length and the adiabaticity has to be found. This was done by performing numerical optimizations with a finite element method (FEM) Maxwell's equation solver (JCMwave). Therefore, the coupler design was parameterized by the lowest possible number of free parameters. Taguchi's method was applied to efficiently optimize the resulting structure. A detailed analysis of this optimization process is provided by Kewes *et al.* [9].

The resulting field distributions for the coupler and the two types of waveguides are presented in FIG. 7.4. For the four free system parameters given in FIG. 7.3b, the following optimized values were calculated: d_{dist} = 80 nm, d_{gap} = 20 nm, d_{arm} = 120 nm, and d_{taper} = 800 nm. It can be seen from the figures that the dielectric field is mainly guided inside the structure, whereas the plasmonic field is located on the lower outer edges of the interface between gold and silica. Thereby, the maximum field density is four times higher as compared to the dielectric part. This shows the strong field enhancement commonly observed in plasmonics. The height gap between the two field distributions is bridged by the tapered coupler section. As shown in FIG. 7.4c, optimum coupling is achieved by allowing the energy to be transferred from the inner section of the V to the outer section of the plasmonic waveguide.

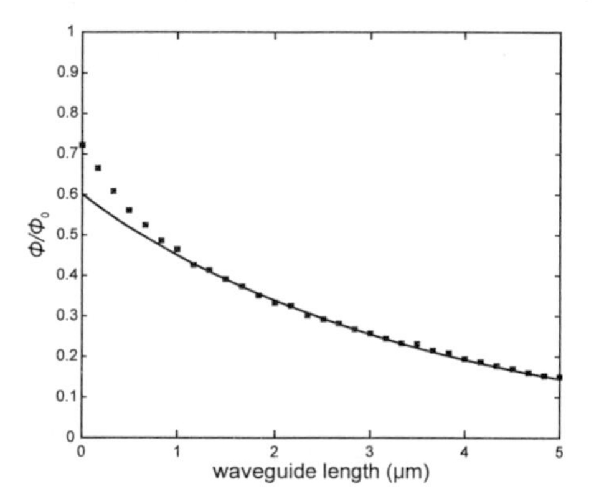

FIG. 7.5. Normalized energy flux at different positions of the plasmonic waveguide (from [9]). The squares are numerical results, whereas the black line is an exponential fit to the data. The intersection of this fit with the flux axis is directly related to the effective coupling efficiency η.

With the optimized design parameters, the effective intermode coupling efficiency can be estimated. Therefore, the effective energy flux was calculated at different waveguide positions z along the total coupler length and fitted by an exponential decay function $f(z) = A_0 \exp(-\alpha z)$ with attenuation constant α. This parameter can be derived from the plasmonic waveguide properties by using a numerical 2D propagation mode solver. By normalizing the energy flux to the incoming fields, the fitting parameter A_0 then directly corresponds to the coupling efficiency at zero waveguide length. As seen in FIG. 7.5, the fitted curve deviates slightly from the calculated flux for short distances. This is due to the influence of scattered light which is introduced when the coupler begins to disturb the field of the dielectric waveguide. As the field becomes more and more converted, the influence of this scattering is reduced and merely simple plasmonic waveguiding is responsible for the calculated energy flux. By this method, a theoretical coupling efficiency η of nearly 60 % could be calculated. The efficiency can be further enhanced to over 70 % in this wavelength regime by using silver instead of gold for the plasmonic structures. However, the results are widely unaffected by the transmitted wavelength as it can be seen from the calculation shown in FIG. 7.6. The presented coupler design shows a broad transmission window in the VIS spectral range and is thus suitable for applications working at multiple wavelengths (plasmon multiplexing). This is in strong contrast to grating couplers which normally show limited bandwidths in the range of a few nanometers. A broad transmission window is also required for ultra-fast plasmonics. In this case, all the higher-order Fourier components of a short pulse must be allowed to pass the plasmonic waveguide.

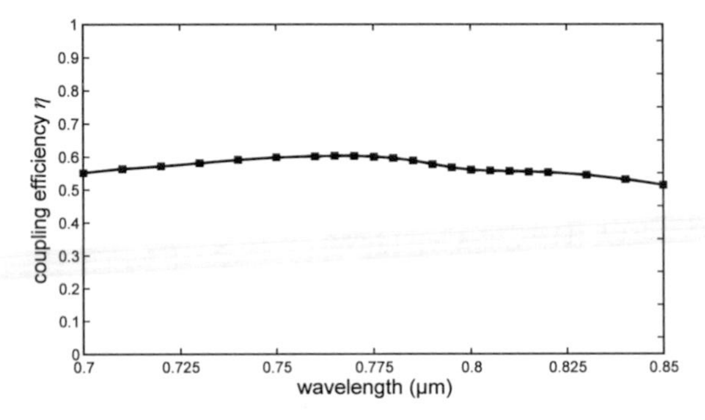

FIG. 7.6. Dependency between the coupling efficiency and the transmitted wavelength in the optimized design (from [9]). A very broad transmission bandwidth is observed which is only limited by the cut-off frequencies of the waveguides.

7.5 Summary and Outlook

In this chapter, alternative waveguide designs for quantum optics applications in the VIS spectral range were presented. Silica offers exceptional transmission properties within this range, but the design of chip-integrated silica waveguide structures is not straightforward (see also Chapter 5) and other materials can be more advantageous in other properties.

By using silicon nitride as waveguide material, direct structuring on top of a substrate becomes possible due to the relatively high refractive index of the material. A possible lower index substrate for nitrides is again silica. It can be used to directly support the written structures in full contact and allows low-loss transmission even for uncladded waveguide structures with strong evanescent fields leaking into the substrate. Another advantage is the possible increase in structure density. This allows a more compact circuit design which may be advantageous for reducing decoherence in later quantum optics applications. As the structuring techniques for nitrides are compatible with standard silicon-based lithographic processing schemes, this material can be used as suitable silica replacement in a wide range of applications. However, nitride materials tend to produce background fluorescence when illuminated with intense laser light. This makes them not suitable for applications were single photons and intense pump fields are routed through a common waveguide structure.

Another possible waveguide design suitable for quantum optics applications is based on plasmonic waveguiding in conductive materials. Within such waveguides, the standard diffraction limit of light can be beaten and ultra-dense chip integration is achievable. The guided light can even be bent on smallest sub-micron radii [27]. However, as such plasmonic waveguides show substantial optical losses, they can only be used on short distances. The effective field density in these plasmonic waveguides can be highly enhanced compared to dielectric structures. This makes them perfectly suitable for processes where interactions between different fields are exploited. Thus, by coupling quantum emitters to such waveguides, the efficiency of mutual coupling can be highly increased. Furthermore, it can be shown that the quantum properties of emitted photons are independent of the guiding mechanism and identical for plasmonic and dielectric waveguides.

The two presented alternative waveguide designs are compatible to each other. They can be combined to allow the realization of highly complex structures consisting of several individual photonic and plasmonic elements. The typically short ranging distances of plasmonic waveguides can be extended by transferring plasmons into guided photons and vice versa. Thus, the high field enhancement or the capabilities of dense integration in plasmonics can be used on small distances, while for connecting or the simple in- and out-coupling to optical fibers a dielectric waveguide can be applied. As all those elements can directly be structured on a common substrate, highly compact and stable quantum circuits can be realized. For the conversion between different fields, most efficient coupler designs are fundamental.

The photon-to-plasmon converter presented within this chapter is specifically designed for such applications. It is based on a simple easy-to-fabricate design which can be fully optimized by using 3D numerical simulations. By setting appropriate design constraints for dielectric single mode transmission at wavelengths around 780 nm, a coupling efficiency of over 60 % was calculated. The presented design uses a V-shaped adiabatic mode transfer for converting photons into plasmons. The results show a very broad transmission bandwidth for the coupler. This allows using it on several different wavelengths without noticeable efficiency drops. The presented coupler design is completely uncladded and thus it allows access to the fields on all parts of the design. This is required for using such integrated elements for future experimental realizations in quantum optics and plasmonics. However, a protection of the guided fields can be realized by simply cladding the structure with a low index material.

[1] J. Bauters, M. Heck, D. John, D. Dai, M. Tien, J. S. Barton, A. Leinse, R. Heideman, D. J. Blumenthal, and J. E. Bowers, "Ultra-low-loss high-aspect-ratio Si_3N_4 waveguides," *Opt. Express*, vol. 19, no. 4, pp. 3163–3174, 2011.

[2] A. Gorin, A. Jaouad, E. Grondin, V. Aimez, and P. Charette, "Fabrication of silicon nitride waveguides for visible-light using PECVD: a study of the effect of plasma frequency on optical properties," *Opt. Express*, vol. 16, no. 18, pp. 13509–13516, 2008.

[3] F. Karouta, K. Vora, J. Tian, and C. Jagadish, "Structural, compositional and optical properties of PECVD silicon nitride layers," *J. Phys. D. Appl. Phys.*, vol. 45, no. 44, p. 445301, 2012.

[4] S. Romero-García, F. Merget, F. Zhong, H. Finkelstein, and J. Witzens, "Silicon nitride CMOS-compatible platform for integrated photonics applications at visible wavelengths," *Opt. Express*, vol. 21, no. 12, pp. 14036–14046, 2013.

[5] W. Barnes, A. Dereux, and T. Ebbesen, "Surface plasmon subwavelength optics," *Nature*, vol. 424, no. 6950, pp. 824–30, 2003.

[6] M. I. Stockman, "Nanoplasmonics: past, present, and glimpse into future," *Opt. Express*, vol. 19, no. 22, pp. 22029–22106, 2011.

[7] D. E. Chang, A. S. Sørensen, E. A. Demler, and M. D. Lukin, "A single-photon transistor using nanoscale surface plasmons," *Nat. Phys.*, vol. 3, no. 11, pp. 807–812, 2007.

[8] A. W. Schell, R. Henze, G. Kewes, and O. Benson, "Photon-to-Plasmon Coupler," US 13/770.157, 2013.

[9] G. Kewes, A. W. Schell, R. Henze, R. S. Schönfeld, S. Burger, K. Busch, and O. Benson, "Design and numerical optimization of an easy-to-fabricate photon-to-plasmon coupler for quantum plasmonics," *Appl. Phys. Lett.*, vol. 102, no. 5, p. 051104, 2013.

[10] M. Barth, N. Nüsse, J. Stingl, B. Löchel, and O. Benson, "Emission properties of high-Q silicon nitride photonic crystal heterostructure cavities," *Appl. Phys. Lett.*, vol. 93, no. 2, p. 021112, 2008.

[11] S. Maier, *Plasmonics: Fundamentals and Applications*. New York: Springer, 2007.

[12] W. Barnes, "Surface plasmon–polariton length scales: a route to sub-wavelength optics," *J. Opt. A Pure Appl. Opt.*, vol. 8, no. 4, pp. 87–93, 2006.

[13] P. Berini, "Plasmon-polariton waves guided by thin lossy metal films of finite width: Bound modes of asymmetric structures," *Phys. Rev. B*, vol. 63, no. 12, p. 125417, 2001.

[14] S. T. Koev, A. Agrawal, H. J. Lezec, and V. A. Aksyuk, "An Efficient Large-Area Grating Coupler for Surface Plasmon Polaritons," *Plasmonics*, vol. 7, no. 2, pp. 269–277, 2011.

[15] R. Yang and Z. Lu, "Subwavelength Plasmonic Waveguides and Plasmonic Materials," *Int. J. Opt.*, vol. 2012, p. 258013, 2012.

[16] G. Wang, H. Lu, X. Liu, Y. Gong, and L. Wang, "Optical bistability in metal-insulator-metal plasmonic waveguide with nanodisk resonator containing Kerr nonlinear medium," *Appl. Opt.*, vol. 50, no. 27, pp. 5287–5290, 2011.

[17] N. Nozhat and N. Granpayeh, "Analysis of the plasmonic power splitter and MUX/DEMUX suitable for photonic integrated circuits," *Opt. Commun.*, vol. 284, no. 13, pp. 3449–3455, 2011.

[18] A. V Akimov, A. Mukherjee, C. L. Yu, D. E. Chang, A. S. Zibrov, P. R. Hemmer, H. Park, and M. D. Lukin, "Generation of single optical plasmons in metallic nanowires coupled to quantum dots," *Nature*, vol. 450, no. 7168, pp. 402–406, 2007.

[19] R. W. Heeres, S. N. Dorenbos, B. Koene, G. S. Solomon, L. P. Kouwenhoven, and V. Zwiller, "On-chip single plasmon detection," *Nano Lett.*, vol. 10, no. 2, pp. 661–664, 2010.

[20] R. A. Wahsheh, Z. Lu, and M. A. G. Abushagur, "Nanoplasmonic couplers and splitters," *Opt. Express*, vol. 17, no. 21, pp. 19033–19040, 2009.

[21] R. Yang, R. A. Wahsheh, Z. Lu, and M. A. G. Abushagur, "Efficient light coupling between dielectric slot waveguide and plasmonic slot waveguide," *Opt. Lett.*, vol. 35, no. 5, pp. 649–651, 2010.

[22] J. Tian, S. Yu, W. Yan, and M. Qiu, "Broadband high-efficiency surface-plasmon-polariton coupler with silicon-metal interface," *Appl. Phys. Lett.*, vol. 95, no. 1, p. 013504, 2009.

[23] Y. Song, J. Wang, Q. Li, M. Yan, and M. Qiu, "Broadband coupler between silicon waveguide and hybrid plasmonic waveguide," *Opt. Express*, vol. 18, no. 12, pp. 13173–13179, 2010.

[24] J.-S. Shin, M.-S. Kwon, and S.-Y. Shin, "Design and analysis of a vertical directional coupler between a three-dimensional plasmonic slot waveguide and a silicon waveguide," *Opt. Commun.*, vol. 284, no. 14, pp. 3522–3527, 2011.

[25] Q. Li and M. Qiu, "Structurally-tolerant vertical directional coupling between metal-insulator-metal plasmonic waveguide and silicon dielectric waveguide," *Opt. Express*, vol. 18, no. 15, pp. 15531–15543, 2010.

[26] J. Mu and W.-P. Huang, "Simulation of three-dimensional waveguide discontinuities by a full-vector mode-matching method based on finite-difference schemes," *Opt. Express*, vol. 16, no. 22, pp. 18152–18163, 2008.

[27] D. Dai, Y. Shi, S. He, L. Wosinski, and L. Thylen, "Silicon hybrid plasmonic submicron-donut resonator with pure dielectric access waveguides," *Opt. Express*, vol. 19, no. 24, pp. 23671–23682, 2011.

Chapter 8

Summary and Outlook

In experimental quantum optics, the fundamental properties of light-matter interaction are investigated on the nanometer scale. The basis of these studies is the ability to prepare, control, and measure well defined quantum states of light and solid-state emitter systems. As quantum states are very fragile in nature, perturbation effects like decoherence and general losses can not be avoided. It is an experimental challenge to circumvent the physical restrictions of the combined quantum systems and allow full control over all relevant physical parameters. The key elements in quantum optics are suitable light sources, an efficient way to store and control single photons, and highly sensitive detector systems. In this thesis, two of the most important components for storing and controlling were investigated. These are namely the optical whispering gallery mode (WGM) microresonators and different types of optical waveguides.

Among the resonators, various cavity systems were analyzed for their applicability in modern quantum optics. In addition to the pure optical properties of such systems, the main fabrication issues and the flexibility in implementing complex experimental setups were investigated. A very important aspect was the possibility to precisely tune the occurring WGM resonances. As for most of the quantum emitters low temperature operation is required, a lot of attention was taken to that point.

The first intensively examined resonator system were novel hollow microspheres also known as microbubbles. Their optical properties are similar to the optical properties of conventional microspheres which have been used in quantum optics for a very long time. As for these, the production process is mainly carried out manually by the thermal processing of thin glass tips. However, the final geometrical and optical properties of such microresonators can be defined in advance only to a very limited extent and the results vary widely among different samples. For manufacturing, a laser-based method was applied and further developed. The resulting resonators are produced with an outstanding reproducibility and in highly symmetric shapes.

The actual microbubbles are manufactured by locally heating and softening a tapered glass capillary within the focus of two CO_2 laser beams. The heated element becomes then expanded by applying an internal overpressure to the capillary. As this purely pressure-induced reversible plastic deformation of the spherical glass body is also possible beneath the specific glass transition temperature, the method has been further investigated as a possible method for tuning the internal WGM resonances in such microbubbles. A direct linear relation between the internally applied pressure and the measured resonance shift, which can be as large as the free spectral range (FSR) of the system, was observed for the examined spheres. This strict linear behavior could be verified by applying the theory of elasticity to the problem. Based on this model, also a simplified relation between microscopically measured and quantified geometric parameters of the microbubbles and their final tuning behavior could be found.

The simple predictability of the final tuning properties in microbubbles already during manufacturing, in combination with the achievable wide tuning ranges, outweighs the disadvantages of manual resonator production to a large extent. In contrast to an artificial photonic molecule consisting of several rigid microspheres, with this tuning technique even multiple microbubble cavities can be combined with precisely tuned or matched individual resonances. Furthermore, cryogenic applicability of microbubbles could be verified by the experiments. In addition to the thermal detuning of the WGM resonances by cooling the system from room temperature to cryogenic temperatures, the capability of pressure tuning under cryogenic conditions could also be demonstrated with helium as inert tuning gas.

Hollow spherical microresonators are therefore a good alternative to conventional spherical microresonator systems, and like these, they can be applied in the context of quantum optics. However, their overall applicability is considerably extended through the presented possibility of WGM resonance tuning by aerostatic pressure. Hollow microresonators are opening countless of applications for sensor and measurement technologies. By giving full access to the interior of the spheres, the evanescent fields of the WGMs can interact with their respective environments on both sides of the walls. This enables completely new possibilities for further optical applications, too.

However, spherical microresonators must subordinate to their directly chip-integrated counterparts in terms of a possible expandability to more complex quantum optical systems. Integrated microresonators can be fabricated by fully automated processes. Even large volumes can be produced within a few production steps. Modern methods of semiconductor technology or cognate techniques can be applied. Thus, elements can be produced with highest accuracy and in best optical quality. Within this work, three different integrated systems were examined. This involves disk-type microresonators made from silica or polymer as well as toroidal microresonators also made from silica.

The two silica-based resonator systems are essentially a simple transfer of the known solid and hollow microsphere resonators to a planar silicon substrate. In addition to adjusting the production process for these structures, other problems, like general chip handling, or coupling and guiding light along the surface of microchips, arise. As for the spherical microresonators, again the problem of precisely tuning the occurring WGM resonances has to be solved. Despite the applied high-precision semiconductor processing techniques, the achievable level of precision is not sufficient for a direct application of the produced structures in quantum optics. Even for such integrated systems, highly accurate methods for tuning the WGM resonances are required. However, these methods are often not suitable for cryogenic applications or the possible tuning ranges are not sufficient for setting the required resonances.

It could be shown that for silica-based integrated microresonators the combination of manual post-production processing by selective etching and thermal fine-tuning by local temperature changes allows a sufficient WGM tuning accuracy for most quantum optics applications. For disk-type silica microresonators, exact resonance tuning with MHz precision over several 100 GHz has been demonstrated within the telecom wavelength range. The presented etching method had a negligible influence on the optical properties of the microresonators and is therefore perfectly suited for refining the manufacturing process with directly applied optical characterization in an everyday laboratory environment. As the applied etching technique primarily works by pure chemical polishing, the presented tuning method is expected to be directly transferable to other substrate-based resonator designs and material systems. The subsequently applied temperature-based fine-tuning process is highly universal and can also be transferred to other platforms.

The applicability of chip-based silica microresonators to cryogenic applications was demonstrated in the context of a proof-of-principle quantum optics experiment. A small number of single-photon emitters were selectively attached to the edge of a toroidal silica microresonator and mutual interaction between the two resonant systems was measured. In addition, in a further experiment the integration of such quantum emitters directly inside the resonators could be demonstrated. These resonators were produced on the basis of a functionalized polymer by means of a two-photon absorption-based laser direct writing scheme. A flexible and truly three-dimensional (3D) production process could be realized with this method. This significantly enhances the design capabilities compared to a conventional two-dimensional (2D) structuring on wafer surfaces. In addition, with this method also novel optical elements for evanescently coupling light to the resonators could be demonstrated. The applicability of such integrated and functionalized polymer-based microresonator systems under cryogenic conditions has also been demonstrated.

As mentioned above, for all chip-based microresonator systems the particular question of efficiently coupling light in and out of theses structures has to be addressed. This applies both to the internal connection of individual structures located on the same microchip as well as for connecting the microchip to other external components. At solid and hollow spherical microresonators, fiber tapers are mainly used for the evanescent coupling. This is not practicable on chip-scale, especially for more complex chip-based future quantum photonic networks. For this purpose, integrated waveguide structures, which can directly be fabricated together with microresonators and other integrated optical components, are required. However, in silica-based material systems the question for an appropriate technology is still matter of research. Silica is, due to its exceptional optical properties, the material of choice in quantum optics, but the relatively low refractive index of that material makes a direct patterning on silicon or other substrates extremely difficult.

In this work, a novel approach for such a direct structuring could be developed and successfully demonstrated. The proposed design is suitable for producing low-loss optical waveguides. The same methods which are used for the fabrication of integrated optical microresonators can be applied for production. Both straight and curved waveguides, as well as one-dimensionally tapered optical coupling elements can be structured. Also extended propagation lengths can be realized with negligible optical losses due to a complete separation between waveguide and substrate. By allowing an appropriate transformation of the mode field diameter, the efficient coupling to external fiber components becomes also possible with these waveguide elements. In addition to presented details about the general fabrication method, within the framework of this thesis also the theoretical background for an appropriate structure design was given and different calculation methods for the modal structure of the devices were compared.

Within this thesis, another waveguide system, which uses a novel direct 3D laser structuring method in diffusion-mediated photopolymers, has been demonstrated. This process is based on a continuously working two-beam laser writing scheme. It allows the dynamic generation of large-scale waveguide structures with variable core sizes. In addition to presenting the method, primarily studies on the resulting mode profiles, the transmission properties, and the maximum possible coupling efficiencies to external optical fibers were presented. It is further possible to directly integrate optical gratings and other holographic microstructures into the polymer. Hence, this system is particularly interesting for the packaging of future quantum photonic microchips as it may replace a variety of different external optical components conventionally used for chip-coupling. Fixed optical connections between multiple microchips, or between a microchip and external optical fibers can be realized. All components can be casted into a common block of polymer. The connecting waveguides can thus directly be patterned from the fiber cores to the edges of the microchip.

In addition to the various waveguide structures, in this thesis also two alternative material systems for integrated applications in quantum optics were presented. By using core materials with a higher index of refraction, they allow to bypass the problem of separating waveguide and substrate by simply structuring a lower index material underneath the guiding core regions. Higher index materials may also be used to allow higher levels of chip-integration. In addition to the reduced geometrical size, especially a potentially higher environmental isolation of realized integrated quantum systems can be implemented due to a reduction of the active chip area. However, as the production requirements rise accordingly, it would be highly speculative to estimate the required amount of process accuracy or if at all a noticeable improvement can be achieved.

As last system in this thesis, an integrated dielectric-plasmonic coupler was presented. These devices are highly interesting for quantum optics applications as they allow a simple experimental access to plasmonics. In addition to the possibility of observing completely new quantum phenomena, a highly significant reduction in the size of the waveguides can be achieved. This is due to a higher field confinement as compared to conventional dielectric waveguides.

The study of single plasmons is much more difficult than the study of individual photons. In such systems, extremely short propagation lengths are commonly observed due to a strong attenuation of the plasmon fields. Furthermore, a direct generation and detection of plasmons is extremely challenging. A possible solution to this problem is a phased approach in which still single photons are generated and detected by means of standard optical methods. For quantum plasmonic applications, these photons have to be converted into plasmons and vice-versa. Therefore, applied quantum plasmonics is often still based on the above-mentioned dielectric methods and extends them with plasmonics.

The said plasmon coupler was especially designed for ease of fabrication. Therefore, merely conventional cleanroom techniques were allowed to be implemented in the processes. Unlike in many other scientific publications of this area, a simple design approach with only four free parameters allows a full 3D numerical simulation of the entire structure. It was successfully shown that highest possible conversion efficiencies can be achieved with the presented design. Thus, the coupler allows experimental investigations in the context of integrated quantum plasmonics and enables many further applications within this exciting new field of research.

Appendix A

Eigenmode Expansion Method

The eigenmode expansion (EME) method is used to numerically model the propagation of electromagnetic waves inside complex waveguide structures [1], [2]. It is based on a local decomposition of an arbitrary input field into a basic set of eigenmodes. The method is also well-known as mode matching (MM) [3] or bidirectional eigenmode propagation (BEP) method [4]. The EME technique is a linear frequency-domain method. It is in contrast to the widely used beam propagation method (BPM) which relies on a one-directional slowly varying envelope approximation [5]. The BPM method does not support bidirectional calculations and neglects reflections.

The eigenmodes used for the EME method are often derived by numerically solving Maxwell's equations in two-dimensional (2D) cross-sections of the analyzed waveguide structure. The results are fully vectorial, but also semi-vectorial approaches are known [6]. The field calculations can be performed by various numerical techniques. For the used simulation software (Photon Design Fimmprop), a whole set of solvers is available. Beside the effective index method (EIM), the software also provides finite-difference time-domain (FDTD) and finite element method (FEM) solvers. This allows selecting the most appropriate solver design for a given problem. Depending on the geometry of the waveguide or its material properties (dielectric or metallic), a solver with highest performance or accuracy can be chosen.

This allows fully three-dimensional (3D) numerical calculations even on elongated structures, or devices consisting of several dielectric and plasmonic elements. Due to their large memory requirements, such calculations are often not possible by other numerical methods, e.g., with FDTD or FEM. For a simple calculation of the modes in optical fibers or fiber-like resonator systems, pure fiber mode solvers are also available. They are based on cylindrical coordinates and take advantage of the circular symmetry of analyzed structures. By choosing an appropriate solver, the calculation performance can be highly enhanced due to a reduction in the number of required eigenmodes.

In the following section, the fundamentals of the EME method are shortly summarized. The mathematical background of field decomposition is presented and the advantage of using EME for simulating optical propagation in waveguides is demonstrated.

Theory

The EME method functionally bases on the orthogonality relation of optical fields [6]. In waveguide structures with a vanishing refractive index variation Δn_z along the axis of propagation (z-axis), the general solution of Maxwell's equations can be expressed at a specific wavelength λ with a harmonic time dependence of the form $e^{i\omega t}$ and with a propagation constant $\beta = n_z \cdot 2\pi/\lambda$ by

$$\mathbf{E}(x,y,z) = \mathbf{E}(x,y)\ e^{-i\beta z}. \tag{A.1}$$

It is possible to derive a condition for the transversal components of these fields by using the reciprocity theorem. For any two guided modes inside a waveguide system,

$$\mathbf{E}_1 = \mathbf{E}_u(x,y)\ e^{-i\beta_u z} \tag{A.2}$$

and

$$\mathbf{E}_2 = \mathbf{E}_v(x,y)\ e^{-i\beta_v z}, \tag{A.3}$$

this condition is given by

$$\int\int_{-\infty}^{\infty} \mathbf{E}_{t,u} \times \mathbf{H}^*_{t,v}\ dxdy = 0,\ \beta_v \neq \beta_w. \tag{A.4}$$

The equation ensures that arbitrary waveguide modes with different propagation constants $\beta_{m,n}$ are always orthogonal to each other. A similar expression can be found for the evanescent fields,

$$\mathbf{E}(x,y,z) = \mathbf{E}_w(x,y)\ e^{-\alpha_w z}. \tag{A.5}$$

Thus, for arbitrary solutions $\mathbf{\Psi} = \{\mathbf{E},\mathbf{H}\}$ of Maxwell's equations in a given waveguide system, it is always possible to find a set of suitable eigenmodes for a series expansion. All possible eigenmodes of that system form a complete orthonormal basis. With this basis, the electromagnetic fields of the propagating and the evanescent waves can be fully described.

For a given cross-sectional plane (z = const.), an arbitrary field can be expressed as superposition of eigenmodes by

$$\mathbf{E}_t(x,y,z)=\sum_m (a_m(z)+a_{-m}(z))\hat{\mathbf{E}}_{t,m}(x,y) \tag{A.6}$$

and

$$\mathbf{H}_t(x,y,z)=\sum_m (a_m(z)+a_{-m}(z))\hat{\mathbf{H}}_{t,m}(x,y). \tag{A.7}$$

Here, a_m and a_{-m} are the expansion coefficients for the forward and backward propagating waves. After normalizing the fields by its power flow, the expansion coefficients for the propagating waves are found as

$$a_{\pm m}(z)=\int\int_{-\infty}^{\infty}\left(\mathbf{E}_t\times\hat{\mathbf{H}}^*_{t,m}\pm\hat{\mathbf{E}}^*_{t,m}\times\mathbf{H}_t\right)_z dxdy. \tag{A.8}$$

For the evanescent fields, they are given by

$$ia_{\pm m}(z)=\int\int_{-\infty}^{\infty}\left(\mathbf{E}_t\times\hat{\mathbf{H}}^*_{t,m}\mp\hat{\mathbf{E}}^*_{t,m}\times\mathbf{H}_t\right)_z dxdy. \tag{A.9}$$

By using Maxwell's equations, the longitudinal fields can then directly be calculated from the transversal field components by

$$\nabla_t\times\mathbf{E}_t=-i\omega\mu_0\mathbf{H}_z \tag{A.10}$$

and

$$\nabla_t\times\mathbf{H}_t=i\omega\varepsilon\varepsilon_0\mathbf{E}_z. \tag{A.11}$$

Hence, full 3D vector fields can be expressed by the expansion coefficients of the eigenmodes. They provide a rigorous solution of Maxwell's equations in non-isotropic linear media. The accuracy of the method is limited by the number of considered eigenmodes. However, it is always possible to estimate the residual error by comparing the two field representations. In most cases, a relatively small number of eigenmodes ($N\approx 10$) is sufficient to approximate field configurations with high accuracy.

The z-dependence of the expanded fields can be expressed by

$$a_{\pm m}(z)=a_{\pm m}(z=0)e^{\mp i\beta_m z} \tag{A.12}$$

and

$$ia_{\pm m}(z)=ia_{\pm m}(z=0)e^{\mp\alpha_m z}. \tag{A.13}$$

For a straight waveguide with constant diameter, the series expansion of the transversal components is z-independent and can be calculated in a single reference frame. However, the final z-dependence of the expansion coefficients shows an exponential behavior.

The EME method can also be used for structures with discontinuities along the axis of propagation [7]. In this case, the field decomposition has to be performed on both sides of a junction. By applying appropriate boundary conditions, an overlap integral can be calculated between the two modes $\mathbf{\Psi}_i$ and $\mathbf{\Psi}_j$. The corresponding coupling coefficients are then given by

$$c_{ij} \propto \int\int_{-\infty}^{\infty} \left(\mathbf{E}_{t,i} \times \mathbf{H}_{t,j}\right)_z dxdy. \tag{A.14}$$

The coupling of different fields is often described by scattering matrices representing the energy transfer through interfaces. They allow a direct mapping between input and output modes propagating in the forward and backward direction, and considering their possible couplings into reflected modes. The scattering matrices can be written in the form $\{\underline{\mathbf{T}}^l, \underline{\mathbf{T}}^r, \underline{\mathbf{R}}^l, \underline{\mathbf{R}}^r\}$ with transmission $\underline{\mathbf{T}}$ and reflection $\underline{\mathbf{R}}$ submatrices for the waves coming from the left l and right r side of the junction. They are defined by the expansion coefficients of the output fields which are expressed in terms of the input fields at the interface by

$$\mathbf{a}_+^r = \underline{\mathbf{T}}^l \mathbf{a}_+^l + \underline{\mathbf{R}}^r \mathbf{a}_-^r, \; \mathbf{a}_-^l = \underline{\mathbf{R}}^l \mathbf{a}_+^l + \underline{\mathbf{T}}^r \mathbf{a}_-^r. \tag{A.15}$$

It is also possible to derive scattering matrices for tapered sections or other modulated waveguide systems by discretizing continuously varying structures. Multiplying the resulting matrices and applying the corresponding z-dependencies to the expansion coefficients allows a direct mapping of the energy transfer between a specific input mode and several output modes, and vice-versa.

The method of a discretized development in system eigenmodes also allows analyzing coupled waveguide structures with varying distances between their components [8]. After evaluating the corresponding eigenmodes of the hybrid system, the energy transfer between the different waveguides can be calculated via the scattering matrices. This technique allows a highly efficient modeling of tapered structures as well as full numerical investigations on the coupling of such elements to other straight or curved waveguide designs. As the series expansion is determined in distinct planes along the direction of propagation, the numerical calculation effort is widely unaffected by the number of elements within these planes. Hence, the calculation effort for a single tapered waveguide is comparable to the calculation effort for a bi-tapered directional coupler as long as the underlying simulation domains are of equal size.

The EME method is a very efficient approach to investigate coupler designs with large dimensions. It is the only method which allows a complete trace of the modal energy transfer and enables to engineer specific mode coupling properties for such devices.

[1] S.-T. Peng and A. A. Oliner, "Guidance and Leakage Properties of a Class of Open Dielectric Waveguides: Part I - Mathematical Formulations," *Trans. Microw. Theory Tech.*, vol. 29, no. 9, pp. 843–855, 1981.

[2] A. Sudbo, "Film mode matching: a versatile numerical method for vector mode field calculations in dielectric waveguides," *Pure Appl. Opt.*, vol. 2, no. 3, pp. 211–233, 1993.

[3] G. V. Eleftheriades, A. S. Omar, L. P. B. Katehi, and G. M. Rebeiz, "Some important properties of waveguide junction generalized scattering matrices in the context of the mode matching technique," *Trans. Microw. Theory Tech.*, vol. 42, no. 10, pp. 1896–1903, 1994.

[4] J. Petracek and J. Luksch, "Bidirectional eigenmode propagation algorithm for 3D waveguide structures," in *13th International Conference on Transparent Optical Networks (ICTON)*, 2011, pp. 1–4.

[5] D. Yevick and B. Hermansson, "New formulations of the matrix beam propagation method: application to rib waveguides," *J. Quantum Electron.*, vol. 25, no. 2, pp. 221–229, 1989.

[6] D. F. G. Gallagher and T. P. Felici, "Eigenmode expansion methods for simulation of optical propagation in photonics: pros and cons," in *Proceedings of SPIE*, 2003, vol. 4987, pp. 69–82.

[7] T. P. Felici and D. F. G. Gallagher, "On propagation through long step tapers," *Opt. Quantum Electron.*, vol. 33, no. 4–5, pp. 399–411, 2001.

[8] M. Galarza, D. Van Thourhout, R. Baets, and M. Lopez-Amo, "Compact and highly-efficient polarization independent vertical resonant couplers for active-passive monolithic integration," *Opt. Express*, vol. 16, no. 12, pp. 8350–8358, 2008.

Appendix B

ECDL System

A tunable grating stabilized external cavity diode laser (ECDL) system was built for probing the narrow-linewidth resonances of whispering gallery mode microresonators. This system was also required for the resonant excitation of nitrogen vacancy defect centers in nanodiamonds [1]. Therefore, a standard single mode laser diode with a central wavelength of 638 nm was combined with a blazed reflection grating to form a Littrow configuration for the external cavity. With this system, an overall tuning range of 5 nm and an effective spectral linewidth below 10 MHz could be achieved. A tunable mode-hop-free scanning range of 15 GHz and stabilized output powers around 1 mW could be successfully demonstrated.

In the following section, this home-build laser system is presented in detail. After a short introduction into the basics of ECDL systems, the specific design of the laser is shown. The system is then characterized by different spectroscopic methods and the results are compared to a commercial laser system.

Theory

The emission of a laser diode is defined by the properties of the cavity and the gain medium. Within these volumes, optical feedback takes place [2]. The cavity is directly built by the semiconductor material or an external resonator, which constantly feeds energy back into the active zone of the diode, is used. The latter configuration is often referred to as ECDL system. If the external cavity resonance has a narrower linewidth than the natural diode emission, the laser light is spectrally filtered by the cavity and then back-coupled into the gain region. This causes the diode, due to internal mode competition, to emit solely at the narrowed cavity resonance. This effect can also be exploited to tune the laser diode. By shifting the spectral properties of the cavity, the emitted laser line can be shifted as well.

For spectral filtering and optical feedback, diffraction gratings can be applied. In a Littrow configuration, the first order is directly back-reflected into the laser diode. For coupling light out of the laser, the zeroth order of diffraction is used. The feedback efficiency can be enhanced by using gratings with blaze angles corresponding to the first order of diffraction. The spectral properties of such a blazed grating can be deduced from the standard grating equation [3],

$$\lambda = 2g\sin\vartheta \,. \tag{B.1}$$

Here, λ is the wavelength of the incident light, ϑ is the angle of diffraction in relation to the grating normal (incident and reflected angles are equal), and g is the grating constant describing the spacing of single grating lines.

The optical feedback path and the diode define a hybrid cavity with length L. This causes the laser to emit in the m-th longitudinal mode at the specific wavelength

$$\lambda_m = \frac{L}{2m} \,. \tag{B.2}$$

With this equation, a free spectral range (FSR) can be defined. This value describes in which spectral range the laser can emit at a specific longitudinal mode order. By using Eq. (B.2), the FSR can be found as

$$\lambda_{FSR} = \frac{\lambda^2}{2L+\lambda} \approx \frac{\lambda^2}{2L} \,. \tag{B.3}$$

The given Eq. (B.1) directly relates the angles of incidence and reflection to the emitted laser wavelength. This wavelength can be controlled by tilting the grating. As such a tilt does not change the effective cavity length L, a specific longitudinal mode order m can be sustained and continuous wavelength tuning is possible. This effect allows shifting the laser frequency around a central wavelength λ_m within the FSR [4]. Hence, it also defines the available tuning range for mode-hop-free laser scanning.

The given equations are strictly valid only in case of perfectly transmitting diodes. Hence, antireflection coatings are normally applied to ECDL laser diodes. If parasitic internal reflections are observed, chaotic multimode behavior may occur [4]. However, if the feedback from the ECDL cavity is strongly dominating, a stable single mode laser emission can still be achieved even with non-antireflection coated laser diodes.

System Design

For the ECDL system, a commercial laser diode without a special anti-reflection coating (Opnext HL6358MG-A) was used. The diode emitted in a single longitudinal mode around a central wavelength of 639 nm. However, the gain range of the diode showed a typical full width at half maximum (FWHM) of 10 nm. The laser diode is specified with a maximum output power of 12 mW. A low-noise power supply for high power diodes (Schäfter + Kirchoff SK9722C+33) was used as current controller. This laser driver directly incorporates two closed-loop PID systems. They allow stabilizing the output power of the ECDL and gives control over the temperature of the diode. The corresponding laser collimator (Schäfter + Kirchoff 48TA-X) was equipped with two suitable thermo-electric coolers (TEC) and an integrated temperature sensor.

For stabilizing the emission frequency, a ruled reflection grating with 1200 lines/mm was used (Thorlabs GR50-1205). A reflection efficiency of around 75 % was observed for linearly polarized light with a wavelength of 638 nm. These gratings are blazed for a wavelength of 500 nm which corresponds to a Littrow angle of 17° 27' in reflection.

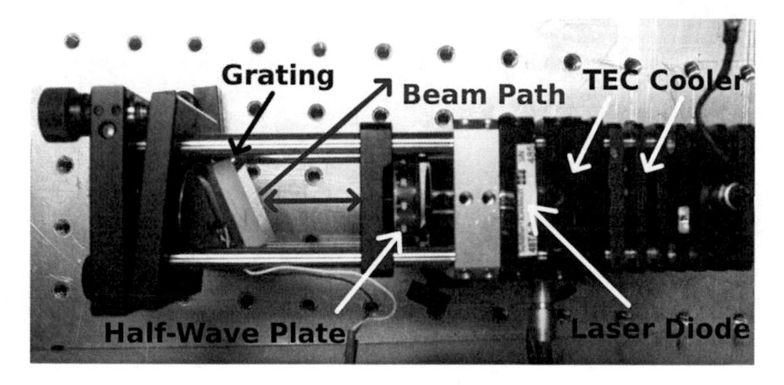

FIG. B.1. Laser head and reflection grating in a Littrow configuration cage assembly.

The laser system was mounted inside a standard cage assembly (Thorlabs). In such a configuration, a stable and warp resistant alignment of the optical components is ensured and high mechanical long-term stability can be achieved. The laser collimator was fixed inside the cage. The reflection grating was hold by a kinematic mirror mount to allow a rough pre-alignment of the Littrow configuration. For fine-tuning, the grating angle was electronically controlled by a piezo actuator. The elements build an effective ECDL cavity length of $L = 7$ cm (see FIG. B.1). This length corresponds to an FSR of roughly 0.003 nm or 2 GHz in the frequency space.

For controlling the polarization of the laser in relation to the grating, a half-wave plate was installed into the beam path. It could be used to rotate the polarization of the emitted laser light. For the suppression of any external feedback, an optical isolator (Thorlabs IO-3D-633-PBS) was used before the laser light is coupled into a fiber.

Characterization

For the characterization of the presented ECDL system, different measurements were performed. These measurements included the maximum achievable output power, the side mode and background suppression, the available tuning ranges for mode-hop-free and coarse scanning, the spectral linewidth, and the long-term stability of the emission.

The maximum output power of the ECDL system was measured by a standard power meter. For the free running diode a peak power of 10 mW was observed. The maximum output power after grating stabilization was 1 mW.

The available tuning ranges of the diode were directly measured by a wavemeter (High Finesse Ångstrom-WS/6). With this technique, the prospected gain range of the free running laser diode could be validated. A coarse overall tuning range of nearly 15 nm was measured. Any potential mode-hops could directly be determined during laser tuning via the real-time frequency control capability of the used wavemeter. They are observed as power discontinuities in an elsewhere continuous laser frequency scan.

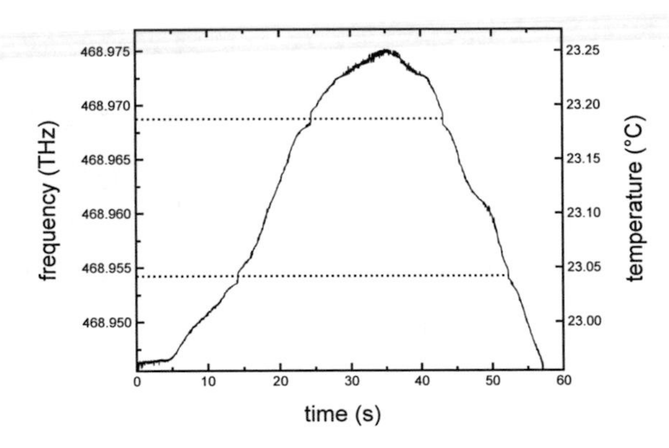

FIG. B.2. The presented ECDL system allows tuning the laser frequency by changing the temperature of the laser diode. With this technique, a possible mode-hop-free tuning range of nearly 15 GHz could be observed.

The presented ECDL system allows two different tuning mechanisms. For high-speed tuning, the tilt of the reflection grating can be controlled by the integrated piezo actuator. However, in a Littrow configuration, this kind of tuning also changes the pointing direction of the output beam. The directional change often causes wavelength dependent power fluctuations when coupling the laser into fibers or subsequent optical components. Another option for controlling the laser frequency is by changing the temperature of the laser diode via the integrated TEC. This process only allows for slow tunings, but keeps the direction of the output beam unaffected. However, both methods showed a sufficient mode-hop-free tuning range of around 15 GHz., The corresponding wavemeter measurement for temperature tuning is shown in FIG. B.2. Mode-hops are visible as vertical lines within the measured frequency curve.

The side mode and background suppression was measured with a grating spectrometer (ACTON SpectraPro 2500i). The results are shown in FIG. B.3. There, the unstabilized spectrum of the free running diode is compared to the grating stabilized spectrum. The observed suppression is enhanced by more than 3 dB due to the stabilization effect.

The linewidth of the laser system was analyzed with a heterodyne measurement in comparison to a commercial grating-stabilized ECDL system (New Focus Velocity). The short-term linewidth of this laser is specified with below 300 kHz. Both lasers were tuned to spectrally overlap their outputs such that a frequency reference for the measured system was defined. The combined laser emission was superimposed on a fast 100 MHz photodetector. The resulting beat signal was measured with an electronic spectrum analyzer (HP 8591a). From these measurements, an upper value of 10 MHz could be estimated for the spectral linewidth of the presented ECDL system.

The values achieved with this system are comparable to other commercial ECDL reference systems. Although the measured spectral linewidth is not as narrow as for the New Focus Velocity, the other investigated optical properties can well compete with commercial ECDL systems (Sacher Lasertechnik Lynx, New Focus Velocity).

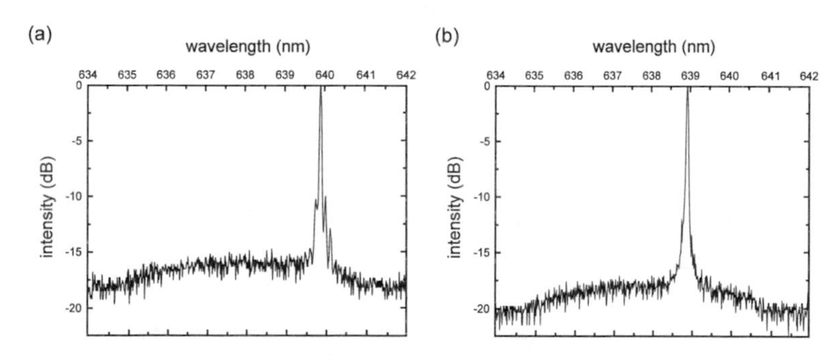

FIG. B.3. Spectral measurement of the (a) free running and (b) grating stabilized laser diode (T = 25 °C, I_D = 34.5 mA). It can be seen that the side mode and background suppression in grating stabilized operation is at least 3 dB higher as for the free running diode.

[1] T. Tyborski, “Aufbau eines schmalbandig-durchstimmbaren gitterstabilisierten Diodenlasersystems,” Humboldt-Universität zu Berlin, 2009.

[2] D. O. North, “Theory of modal character, field structure, and losses for semiconductor lasers,” *J. Quantum Electron.*, vol. 12, no. 10, pp. 616–624, 1976.

[3] H. Noda, T. Namioka, and M. Seya, “Geometric theory of the grating,” *J. Opt. Soc. Am.*, vol. 64, no. 8, p. 1031, 1974.

[4] J. Sacher, D. Baums, P. Panknin, W. Elsässer, and E. O. Göbel, “Intensity instabilities of semiconductor lasers under current modulation, external light injection, and delayed feedback,” *Phys. Rev. A*, vol. 45, no. 3, pp. 1893–1905, 1992.

Appendix C

Photon Statistics and Measurement

In quantum optics, the classical electromagnetic fields are quantized and described by operator fields [1]. In this representation, the corresponding vector fields or potentials are Fourier expanded in a complete set of basis wave functions which describe the physical eigenstates of a system. The corresponding field coefficients can be written in the form of creation and annihilation operators $\{\hat{a}^{\dagger}, \hat{a}\}$. For example, in a box with volume $V = L^3$ the electric field operator $\hat{\mathbf{E}}(\mathbf{r})$ can be written as [2]

$$\hat{\mathbf{E}}(\mathbf{r}) = i\sum_{\mathbf{k},\mu}\sqrt{\frac{\hbar\omega_{\mathbf{k}}}{2V\varepsilon_0}}(\hat{a}_{\mathbf{k}}^{(\mu)}\mathbf{e}^{(\mu)}e^{i\mathbf{k}\cdot\mathbf{r}} - \hat{a}_{\mathbf{k}}^{\dagger(\mu)}\mathbf{e}^{*(\mu)}e^{-i\mathbf{k}\cdot\mathbf{r}})\,\text{with}\,\mu \in \{1,-1\}. \tag{C.1}$$

Here, $\hbar$ is Planck's constant, ε_0 is the dielectric permittivity, $\mathbf{k}$ is the wave vector, and ω is the corresponding angular frequency. The two complex polarization vectors $\mathbf{e}^{(\mu)}$ are normalized and chosen perpendicular to the wave vector. This representation of the fields is similar to the well-known ladder operators in a quantum harmonic oscillator; it describes a field consisting of discrete energy packets called photons. They are carriers of energy, momentum, and spin.

Photons are bosonic and thus several photons can occupy a common energy state. The number of photons in a specific mode can be found be simply applying the number operator $\hat{N}_{\mathbf{k}} = \hat{a}_{\mathbf{k}}^{\dagger}\hat{a}_{\mathbf{k}}$ to the operator field. An eigenstate of the number operator is called a photon number state or Fock state, the formalism itself is often referred to as number state representation [3]. In a specific mode, a well-defined energy portion $\hbar\omega$ is carried per photon. The Hamiltonian of a system can thus be described by

$$\hat{H} = \sum_{\mathbf{k},\mu}\hbar\omega(\hat{a}_{\mathbf{k}}^{\dagger(\mu)}\hat{a}_{\mathbf{k}}^{(\mu)} + \frac{1}{2}). \tag{C.2}$$

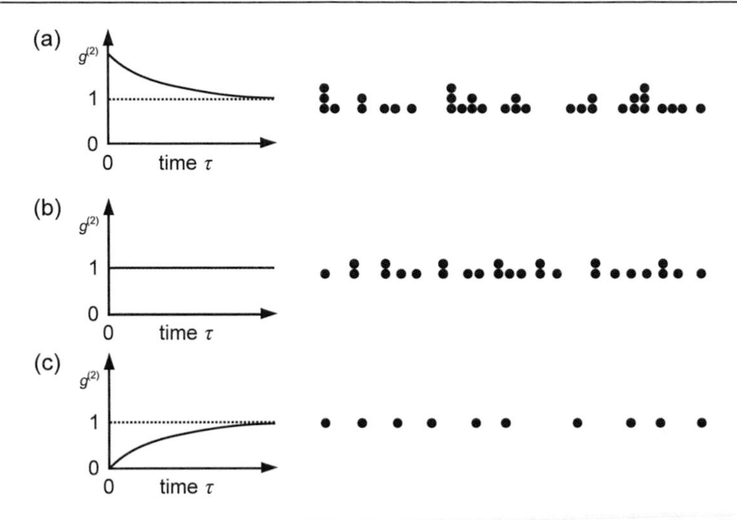

FIG. C.1. Second-order intensity correlation functions of different light sources. (a) Thermal light source with high photon correlation (bunched light). Several photons can occupy the same temporal mode. (b) Laser light source (Poissonian probability distribution). (c) Single photon emitter (antibunched light). There is one photon per temporal mode at maximum.

In time dependent systems, the number of photons is variable and can change over time. Hence, the energy of the system becomes also time dependent and can be used to describe the photon statistics of that system. In the given operator field representation, the variance of the photon number distribution in a specific mode can be calculated from the number operator by [2]

$$V_n = \langle n^2 \rangle - \langle n \rangle^2 = \langle (\hat{a}^\dagger \hat{a})^2 \rangle - \langle \hat{a}^\dagger \hat{a} \rangle^2 . \tag{C.3}$$

With this expression, the classical second-order intensity correlation function $g^{(2)}(0)$ can be calculated from the photon number [4]. With the intensity $I = \mathbf{E} \cdot \mathbf{E}^*$ of the light field, this can be written as

$$g^{(2)}(0) = \frac{V_n - \langle n \rangle}{\langle n \rangle^2} + 1 = \frac{\langle (\hat{a}^\dagger)^2 \hat{a}^2 \rangle}{\langle \hat{a}^\dagger \hat{a} \rangle^2} \cong \frac{\{I(0)^2\}_t}{\{I(0)\}_t^2} . \tag{C.4}$$

The $\{\cdots\}_t$ indicates the time average within a time interval t. This value basically describes the probability of having multiple photons within a mode, normalized by the overall probability of this specific multi-photon state. The more general case of a second-order intensity correlation function $g^{(2)}(\tau)$ includes a time difference $\tau << t$ between two different field configurations. This function is given by

$$g^{(2)}(\tau) = \frac{\left\langle \hat{a}^{\dagger}(0)\hat{a}^{\dagger}(\tau)\hat{a}(\tau)\hat{a}(0) \right\rangle}{\left\langle \hat{a}^{\dagger}\hat{a} \right\rangle^2} \cong \frac{\{I(0)I(\tau)\}_t}{\{I(0)\}_t^2}. \tag{C.5}$$

The $g^{(2)}$-function allows to measure the photon statistics of light. Depending on the source, different regimes can be defined [5]. In the case of $g^{(2)}(0) = 1$, the underlying statistics is Poissonian. An example is the highly coherent light field of a laser. The emitted photons have a completely random spacing. For a classical thermal light source, $g^{(2)}(0) > 1$ can be found which leads to a super-Poissonian distribution. The photons are highly correlated and appear as bunched. The number of fluctuations is in this case larger as for the coherent state. For ideal single photon emitters, like quantum dots or single molecules, only one photon can be emitted at a specific time interval and thus $g^{(2)}(0) < 1$ is measured [6]. The statistics is sub-Poissonian and the photons appear as antibunched. The variance of this distribution is less than its mean value and smaller as for the coherent state. In FIG. C.1, an illustration of the three regimes is shown.

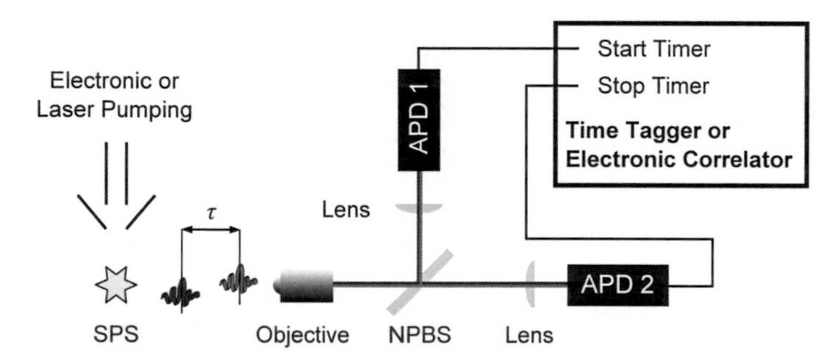

FIG. C.2. Schematic of a typical Hanbury Brown and Twiss (HBT) setup in quantum optics. A continuously pumped single photon source (SPS) emits photons with time difference τ. The photons are collected by a microscope objective and directed by a non-polarizing beam splitter (NPBS) with equal splitting ratio. After detecting a photon on the first avalanche photodiode (APD 1), the trigger signal is electronically routed to an electronic correlator or time tagger which basically measures the time difference between consecutive start and stop photons.

For practically measuring the photon statistics of a non-classical light field, an intensity measurement with photon number resolving photodetectors can be performed [7]. However, except for some specific cases, the dead time of such devices prevent a direct statistics measurement with a single detector [8]. By using two identical photodetectors, this problem can be circumvented. The corresponding experimental setup was first proposed by Hanbury Brown and Twiss (HBT) for measuring spatial correlations in astrophysics [9]. In FIG. C.2 a HBT setup, as it is typically used in quantum optics, is schematically shown. It basically consists of a semi-transparent mirror which splits an incoming photon stream with equal probability. The two detectors are connected to a correlator. For one of the detectors, an additional delay line is included to allow a time difference between simultaneous detection events on both devices. The correlator detects the arrival times of individual photons and calculates the conditional probability to detect a photon in one channel when another photon was previously detected on the other channel. Thus, the influence of the detector dead time can be fully avoided during photon statistics measurements.

[1] R. Loudon, *The Quantum Theory of Light*. Oxford: Oxford University Press, 2000.

[2] M. O. Scully and M. S. Zubairy, *Quantum Optics*. Cambridge: Cambridge University Press, 1997.

[3] S. Schiller, G. Breitenbach, S. Pereira, T. Müller, and J. Mlynek, "Quantum Statistics of the Squeezed Vacuum by Measurement of the Density Matrix in the Number State Representation," *Phys. Rev. Lett.*, vol. 77, no. 14, pp. 2933–2936, 1996.

[4] G. Rempe, R. Thompson, R. Brecha, W. Lee, and H. Kimble, "Optical bistability and photon statistics in cavity quantum electrodynamics," *Phys. Rev. Lett.*, vol. 67, no. 13, pp. 1727–1730, 1991.

[5] R. Loudon, "Non-classical effects in the statistical properties of light," *Reports Prog. Phys.*, vol. 43, no. 7, pp. 913–949, 1980.

[6] P. Michler, A. Imamoglu, M. Mason, P. Carson, G. Strouse, and S. Buratto, "Quantum correlation among photons from a single quantum dot at room temperature," *Nature*, vol. 406, no. 6799, pp. 968–970, 2000.

[7] P. Meystre and M. Sargent, *Elements of Quantum Optics*. Berlin: Springer, 1999.

[8] G. A. Steudle, S. Schietinger, D. Höckel, S. N. Dorenbos, I. E. Zadeh, V. Zwiller, and O. Benson, "Measuring the quantum nature of light with a single source and a single detector," *Phys. Rev. A*, vol. 86, no. 5, p. 053814, 2012.

[9] R. Hanbury Brown and R. Q. Twiss, "Correlation between Photons in two Coherent Beams of Light," *Nature*, vol. 177, no. 4497, pp. 27–29, 1956.

Symbols and Abbreviations

$\mathbf{E}(\mathbf{r})$, $\mathbf{H}(\mathbf{r})$	electric and magnetic vector field
ε_0, μ_0	dielectric permittivity and permeability of the vacuum
I	field intensity
c	vacuum speed of light
t	time
λ_{res}, ν_{res}	resonance wavelength and frequency
n_{eff}, r_{eff}, V_{eff}	effective refractive index, radius, and volume
l, m, n	mode numbers of microresonators: azimuthal mode number, polar mode number, radial mode number
U	field energy
τ	photon lifetime
Q	quality factor
α, β	attenuation and propagation coefficient
k, ω	wave number and angular frequency
$Y_l^m(\vartheta,\varphi)$	Laplace spherical harmonics of l-th degree and m-th order
J_m, Y_m, H_m	first and second kind Bessel and Hankel functions of m-th order
j_l, y_l, h_l	first and second kind spherical Bessel and spherical Hankel functions of l-th order
$\underline{\mathbf{S}}$	scattering matrix
L	cavity length
t_c, κ_c	complex amplitude transmission and coupling coefficient
d	distance or diameter
a, b	outer and inner radius of a microbubble
w	wall thickness or width
p_i, p_0	inner and outer pressure
σ_r, u_r	principal radial stress and radial displacement
υ, G, K	Poisson's ratio, shear and bulk modulus

C	elasto-optic constant
χ	geometric microbubble parameter
T	temperature
α, β	linear thermal expansion and thermo-optic coefficient
h_s	slab height
γ, κ	attenuation coefficient and transverse wave vector
θ_{max}	maximum waveguide acceptance angle
R_a	arithmetic mean surface roughness
$g^{(2)}(\tau)$	time-depended second order intensity correlation function
N	number of photon emitters
$k_{exc/em}$	photon excitation/emission rate
$\hat{a}^{\dagger}, \hat{a}$	creation and annihilation operators
$\hat{\mathbf{E}}(\mathbf{r})$	electric vector field operator
$\hbar$	Planck's constant
$\mathbf{e}$	unity basis vector
$\hat{N}, n$	photon number operator and photon number
V	variance

2D	Two-Dimensional
3D	Three-Dimensional
ARROW	Antiresonant Reflecting Optical Waveguide
APD	Avalanche Photodiode
BHF	Buffered Hydrofluoric Acid
BPM	Beam Propagation Method
CMT	Coupled Mode Theory
CQED	Cavity Quantum Electrodynamics
CVD	Chemical Vapor Deposition
DLW	Direct Laser Writing
DNC	Diamond Nanocrystal
DSR	Differential Shift Rate
DSO	Digital Storage Oscilloscope
ECDL	External Cavity Diode Laser
EIC	Electronic Integrated Circuit
EIM	Effective Index Method
EME	Eigenmode-Expansion-Method
FDTD	Finite-Difference Time-Domain
FEM	Finite Element Method
FHD	Flame Hydrolysis Deposition
FIB	Focused Ion Beam
FSR	Free Spectral Range

FWHM	Full Width at Half Maximum
GFRP	Glass-Fiber Reinforced Plastics
GOPHER	Generating Photonic Elements by Refractive Ion Etching
HBT	Hanbury Brown and Twiss
HCN	Hydrogen Cyanide
HPCVD	High Pressure Chemical Vapor Deposition
IBE	Ion Beam Etching
IR	Infrared
LCORR	Liquid-Core Optical Ring-Resonators
LN	Liquid Nitrogen
LPCVD	Low Pressure Chemical Vapor Deposition
MFD	Mode Field Diameter
NPBS	Non-Polarizing Beam Splitter
NIR	Near-Infrared
N-V	Nitrogen-Vacancy
PCB	Photonic Circuit Board
PDLC	Polymer Dispersed Liquid Crystal
PE	Plasma Etching
PEVCD	Plasma Enhanced Chemical Vapor Deposition
PIC	Photonic Integrated Circuit
Q Factor	Quality Factor
RIE	Reactive Ion Etching
SEM	Scanning Electron Microscope
SOI	Silicon-On-Insulator
SPARROW	Stripe-Line Pedestal Anti-Resonant Reflecting Optical Waveguide
SPP	Surface Plasmon Polariton
SPS	Single Photon Source
STED	Stimulated Emission Depletion
TE	Transverse Electric
TEC	Thermo-Electric Cooler
TIR	Total Internal Reflection
TM	Transverse Magnetic
TPA	Two Photon Absorption
UV	Ultraviolet
VCSEL	Vertical Cavity Surface Emitting Laser
VIS	Visible
WGM	Whispering Gallery Mode
ZPL	Zero Phonon Line

Publication List

The following list of articles (in reverse chronological order) has been published by the author. Work that is marked by an asterisk is not part of this thesis.

Peer-reviewed journals

- Andreas W. Schell, Johannes Kaschke, Joachim Fischer, Rico Henze, Janik Wolters, Martin Wegener, and Oliver Benson. Three-dimensional quantum photonic elements based on single nitrogen vacancy centres in laser-written 3D microstructures. *Nature Scientific Reports* **3**, p. 1577, 2013.

- Günter Kewes, Andreas W. Schell, Rico Henze, Rolf S. Schönfeld, Sven Burger, Kurt Busch, and Oliver Benson. Design and numerical optimization of an easy-to-fabricate photon-to-plasmon coupler for quantum plasmonics. *Applied Physics Letters* **102(5)**, p. 051104, 2013.

- Rico Henze, Christoph Pyrlik, Andreas Thies, Jonathan M. Ward, Andreas Wicht, and Oliver Benson. Fine-tuning of whispering gallery modes in on-chip silica microdisk resonators within a full spectral range. *Applied Physics Letters* **102(4)**, p. 041104, 2013.

- Rico Henze, Jonathan M. Ward, and Oliver Benson. Temperature independent tuning of whispering gallery modes in a cryogenic environment. *Optics Express* **21(1)**, pp. 675-680, 2013.

- Rico Henze, Tom Seifert, Jonathan M. Ward, and Oliver Benson. Tuning whispering gallery modes using internal aerostatic pressure. *Optics Letters* **36(23)**, pp. 4536-4538, 2011.

- *Markus Gregor, Christoph Pyrlik, Rico Henze, Andreas Wicht, Achim Peters, and Oliver Benson. An alignment-free fiber-coupled microsphere resonator for gas sensing applications. *Applied Physics Letters* **96(23)**, p. 231102, 2010.

- Markus Gregor, Rico Henze, Tim Schröder, and Oliver Benson. On-demand positioning of a preselected quantum emitter on a fiber-coupled toroidal microresonator. *Virtual Journal of Nanoscale Science & Technology* **20(18)**, 2009.

- Markus Gregor, Rico Henze, Tim Schröder, and Oliver Benson. On-demand positioning of a preselected quantum emitter on a fiber-coupled toroidal microresonator. *Applied Physics Letters* **95(15)**, p. 153110, 2009.

Patents

- Andreas W. Schell, Rico Henze, Günter Kewes, and Oliver Benson. Photon-to-plasmon coupler. *U.S. patent application* **13/770.157**, 2013.

- Rico Henze, Andreas Thies, and Oliver Benson. Wellenleiteranordnung. *German patent application* **10 2012 222 898.5**, 2012.

Conference Proceedings

- Andreas W. Schell, Tanja Neumer, Qiang Shi, Johannes Kaschke, Joachim Fischer, Rico Henze, Janik Wolters, Martin Wegener, and Oliver Benson, On-chip integration of NV centers in three-dimensional laser-written microstructures for single photon applications. in *Research in Optical Sciences*, p. QTu3B.5, 2014.

- Andreas W. Schell, Tanja Neumer, Qiang Shi, Johannes Kaschke, Joachim Fischer, Rico Henze, Janik Wolters, Martin Wegener, and Oliver Benson. Nanophotonics with single photons from NV centers in three-dimensional laser-written microstructures. in *Frontiers in Optics 2013*, p. FW1C.2, 2013.

- Andreas W. Schell, Johannes Kaschke, Joachim Fischer, Rico Henze, Janik Wolters, Martin Wegener, and Oliver Benson. Single photon nanophotonics using NV centers in three-dimensional laser-written microstructures. in *2013 Conference on Lasers and Electro-Optics*, p. CK_7_1, 2013.

- Andreas W. Schell, Johannes Kaschke, Joachim Fischer, Rico Henze, Janik Wolters, Martin Wegener, and Oliver Benson. Three-dimensional quantum photonic elements based on nanodiamonds in laser-written 3D microstructures. in *Proceedings of SPIE* **8635**, p. 863515, 2013.

- Jonathan M. Ward, Rico Henze, Markus Gregor, Christoph Pyrlik, Andreas Wicht, Andreas Thies, Achim Peters, Síle N. Chormaic, and Oliver Benson. Integrated whispering-gallery mode resonators for fundamental physics and sensing applications. in *Proceedings of SPIE* **8236**, p. 82361C, 2012.

- Michael Barth, Markus Gregor, Rico Henze, Tim Schröder, Nils Nüsse, Bernd Löchel, and Oliver Benson. Hybrid approaches toward single emitter coupling to optical microresonators. in *Proceedings of SPIE* **7579**, p. 757918, 2010.

- *Susanna Orlic, Christian Müller, Enrico Dietz, Sven Frohmann, Jonas Gortner, Bruno Heimke, and Rico Henze. Performance of photopolymer materials for microlocalized volume storage. in *Proceedings of SPIE* **6335**, p. 633518, 2006.

- *Susanna Orlic, Enrico Dietz, Sven Frohmann, Jonas Gortner, Bruno Heimke, Rico Henze, and Christian Müller. Progress in microholographic data storage. in *Proceedings of SPIE* **6335**, p. 633507, 2006.

- *Susanna Orlic, Enrico Dietz, Sven Frohmann, Jonas Gortner, Bruno Heimke, Rico Henze, and Christian Müller. Microholographic data storage: multilayers at the optical resolution limit. in *Proceedings of SPIE* **6187**, p. 618703, 2006.

- *Susanna Orlic, Enrico Dietz, Sven Frohmann, Jonas Gortner, Rico Henze, and Christian Müller. Microholographic data storage at the optical resolution limit. in *2005 Conference on Lasers and Electro-Optics Europe*, p. 152, 2005.

- *Susanna Orlic, Enrico Dietz, Sven Frohmann, Jonas Gortner, Bruno Heimke, Rico Henze, and Christian Müller. Microholographic information storage in photopolymers. in *Proceedings of the International Symposium OPTRO*, 2005.

Contributed Talks

- Andreas W. Schell, Johannes Kaschke, Joachim Fischer, Rico Henze, Janik Wolters, Martin Wegener, and Oliver Benson. NV-centers as single-photon emitters integrated into three-dimensional laser-written micro-structures. *Frühjahrstagung der Deutschen Physikalischen Gesellschaft*, March 18-22, Hannover, Germany, 2013.

- Michael Barth, Markus Gregor, Rico Henze, Günter Kewes, Andreas W. Schell, Stefan Schietinger, Tim Schröder, Janik Wolters, Oliver Benson, Nils Nüsse, Max Schöngen, Bernd Löchel, Henning Döscher, Thomas Hannappel, Tobias Hanke, Alfred Leitenstorfer, and Rudolph Bratschitsch. Fundamental photonic hybrid systems based on defect centers in diamond. *Optics of Excitons in Confined Systems 12*, September 12-16, Paris, France, 2011.

- Rico Henze, Markus Gregor, Tim Schröder, and Oliver Benson. On-demand positioning of a preselected quantum emitter on a fiber-coupled toroidal microresonator. *Frühjahrstagung der Deutschen Physikalischen Gesellschaft*, March 8-12, Hannover, Germany, 2010.

- *Sven Frohmann, Enrico Dietz, Jonas Gortner, Bruno Heimke, Rico Henze, Christian Müller, and Susanna Orlic. Microholographic data storage: multilayers at the optical resolution limit. *Photonics North 2006*, June 5-8, Quebec City, Canada, 2006.

- *Enrico Dietz, Sven Frohmann, Jonas Gortner, Bruno Heimke, Rico Henze, Christian Müller, and Susanna Orlic. Progress and achievements in Microholas project. *COST P8 Meeting*, May 26-27, Loutraki, Greece, 2006.

- *Susanna Orlic, Enrico Dietz, Sven Frohmann, Jonas Gortner, Rico Henze, and Christian Müller. High-density multilayer recording of microgratings for 3D optical data storage. *COST P8 Meeting*, September 16-17, Paris, France, 2005.

- *Susanna Orlic, Enrico Dietz, Sven Frohmann, Jonas Gortner, Bruno Heimke, Rico Henze, Christian Müller. Demonstration and modeling of high density multilayer recording for microholographic data storage. *Optics & Photonics 2005*, July 31-August 4, San Diego, US, 2005.

- *Christian Müller, Enrico Dietz, Sven Frohmann, Jonas Gortner, Bruno Heimke, Rico Henze, and Susanna Orlic. Recent development in microholographic data storage. *Holography 2005*, Varna, Bulgaria, May 22-25, 2005.

- *Enrico Dietz, Sven Frohmann, Christian Müller, Rico Henze, and Susanna Orlic. Dynamisches Schreiben und Lesen im Submikrometerbereich. *Frühjahrstagung der Deutschen Physikalischen Gesellschaft*, March 4-9, Berlin, Germany, 2005.

- *Sven Frohmann, Enrico Dietz, Christian Müller, Rico Henze, and Susanna Orlic. Photopolymere als Speichermaterialien. *Frühjahrstagung der Deutschen Physikalischen Gesellschaft*, March 4-9, Berlin, Germany, 2005.

Posters

- Günter Kewes, Andreas W. Schell, Rico Henze, Sven Burger, and Oliver Benson. Towards quantum plasmonics on a chip: an efficient photon-plasmon-coupler for single-mode operation. *Frühjahrstagung der Deutschen Physikalischen Gesellschaft*, March 25-30, Berlin, Germany, 2012.

- Rico Henze, Tom Seifert, Christoph Pyrlik, Jonathan M. Ward, and Oliver Benson. Optical microresonators. *IONS-11*, February 22-25, Paris, France, 2012.

- Roland Albrecht, Christian Deutsch, Jakob Reichel, Tim Schröder, Rico Henze, Oliver Benson, and Christoph Becher. Towards coupling of a single N-V center in diamond to a fiber based micro-cavity. *Frühjahrstagung der Deutschen Physikalischen Gesellschaft*, March 13-18, Dresden, Germany, 2011.

- Rico Henze, Markus Gregor, Christoph Pyrlik, Andreas Wicht, Achim Peters, and Oliver Benson. Toroidal microresonators as high-tech platform for applications in sensor, analysis, laser measurement and optical telecommunication systems. *Laser Optics Berlin*, March 22-24, Berlin, Germany, 2010.

- Rico Henze, Markus Gregor, Tim Schröder, and Oliver Benson. Cavity-QED with toroidal microresonators. *IONS-6*, July 08-10, Glasgow, UK, 2009.

- Rico Henze, Markus Gregor, Tim Schröder, Helmar Kostial, Edith Wiebicke, and Oliver Benson. Towards controlled cavity-QED experiments with toroidal microresonators. *IONS-5*, February 19-21, Barcelona, Spain, 2009.

- Rico Henze, Markus Gregor, Tim Schröder, Helmar Kostial, Edith Wiebicke and Oliver Benson. Towards controlled cavity-QED experiments with toroidal microresonators. *Frühjahrstagung der Deutschen Physikalischen Gesellschaft*, March 10-14, Darmstadt, Germany, 2008.

Danksagung

An dieser Stelle möchte ich mich bei allen bedanken, die durch ihre großartige Hilfe und Unterstützung ganz wesentlich zum Gelingen dieser Arbeit mit beigetragen haben. Ihr habt meine Promotionszeit zu einer intensiven und unvergesslichen Erfahrung für mich gemacht.

Mein ganz besonderer Dank gilt Prof. Oliver Benson dafür, dass er mir mit einem spannenden und fordernden Thema die Möglichkeit zu einer Promotion in seiner Arbeitsgruppe gegeben hat. Auch für die exzellente fachliche Betreuung dieser Arbeit möchte ich mich herzlich bedanken. Die vielen anregenden Diskussionen sowie seine hilfreichen Ratschläge haben über so manche schwierige Phase hinweggeholfen.

Weiterhin möchte ich mich unbedingt bei Prof. Achim Peters und Andreas Wicht vom Ferdinand-Braun-Institut für die sehr angenehme und erfolgreiche Zusammenarbeit bedanken. Ich habe mich in den Räumen des FBH stets willkommen gefühlt und dabei auch einen Einblick in die Feinheit der Metrologie gewinnen können.

Auf technologischer Seite möchte ich mich vor allem bei Andreas Thies für die viele Reinraumarbeit zur Umsetzung unserer unzähligen Ideen bedanken. Die interessanten Diskussionen dazu werden mir unvergesslich bleiben. Des Weiteren möchte ich hierbei aber auch Edith Wiebicke und Helmar Kostial vom Paul-Drude-Institut danken. Alles was ich jemals über Halbleiterprozessierung gelernt habe, verdanke ich diesen beiden Reinraumexperten. Und für alle sonstigen Fragestellungen stand gleichbedeutend stets Herr Dipl.-Ing. Klaus Palis zur Seite. Erst durch sein umfangreiches elektronisches Fachwissen wurde so manch experimenteller Aufbau überhaupt erst möglich.

Ebenfalls ganz herzlich bedanken möchte ich mich zudem bei Jonathan M. Ward, Markus Gregor und Andreas W. Schell. Auch wenn nicht jedes unserer Experimente von Erfolg gekrönt war, möchte ich die dabei gemachten Erfahrungen niemals missen.

Darüber hinaus bedanke ich mich auch bei Tim Schröder und Christoph Pyrlik sowie bei meinen beiden Bacheloranden Tobias und Tom. Insbesondere die nächtlichen Messreihen im Labor mit Tim und Christoph sowie die vielen angeregten Gespräche werden mir fehlen.

Daneben möchte ich mich auch bei allen Mitgliedern der Arbeitsgruppen NANO, QOM und AMO sowie bei all meinen sonstigen „Physiker-Freunden" bedanken. Neben meinen Zimmerkollegen Alex, Gesine, Otto und Stefan möchte ich dabei besonders Alan, Günter, Ingmar, Jackie, Janik, Johannes, Jonas, Martin, Michael und Moritz erwähnen. Die gemeinsamen Stunden mit euch waren immer ein Erlebnis. Auch unsere gemeinsamen Aktivitäten im Rahmen des Berlin Optik Student Chapters haben mir stets viel Freude bereitet.

Abschließend möchte ich mich sowohl bei meinen Eltern Angelika und Matthias, meinem Bruder Alexander, Wenke, Sandro, René und all den anderen tollen und bisher nicht genannten Menschen in meinem Leben auf das aller herzlichste für ihre Liebe, Zuneigung und Freundschaft bedanken.